2013

Beginning an Era of Hope and Harmony

R. LATAINE TOWNSEND

BALBOA
PRESS
A DIVISION OF HAY HOUSE

Copyright © 2011 R. Lataine Townsend

All rights reserved. No part of this book may be used or reproduced by any means, graphic, electronic, or mechanical, including photocopying, recording, taping or by any information storage retrieval system without the written permission of the publisher except in the case of brief quotations embodied in critical articles and reviews.

Balboa Press books may be ordered through booksellers or by contacting:

Balboa Press
A Division of Hay House
1663 Liberty Drive
Bloomington, IN 47403
www.balboapress.com
1-(877) 407-4847

Because of the dynamic nature of the Internet, any web addresses or links contained in this book may have changed since publication and may no longer be valid. The views expressed in this work are solely those of the author and do not necessarily reflect the views of the publisher, and the publisher hereby disclaims any responsibility for them.

The author of this book does not dispense medical advice or prescribe the use of any technique as a form of treatment for physical, emotional, or medical problems without the advice of a physician, either directly or indirectly. The intent of the author is only to offer information of a general nature to help you in your quest for emotional and spiritual well-being. In the event you use any of the information in this book for yourself, which is your constitutional right, the author and the publisher assume no responsibility for your actions.

Any people depicted in stock imagery provided by Thinkstock are models, and such images are being used for illustrative purposes only.
Certain stock imagery © Thinkstock.

ISBN: 978-1-4525-4343-7 (sc)
ISBN: 978-1-4525-4344-4 (hc)
ISBN : 978-1-4525-4342-0 (e)

Library of Congress Control Number: 2011961676

Printed in the United States of America

Balboa Press rev. date: 12/29/2011

For my father
who had no sons
or nephews
to carry on the family name,
I publish in the name with which I was born.

Contents

1. **INTRODUCTION**   1
   - The Journey   3
2. **THE MAYANS**   9
   - The Ancestors   11
   - Catastrophes   16
   - Atlantis   20
   - The Olmecs   30
   - Izapa, the Intermediate Culture   36
   - The Mayan Culture   40
3. **TIME AND THE MAYAN CALENDAR**   55
   - Cycles   57
   - Precession   59
   - The Calendars   64
   - The Stars   67
4. **ASTRONOMY AND ASTROLOGY**   71
   - Our Sun   73
   - The Milky Way   78
   - Other Dangers   81
   - Effects on Humans   84
   - Cosmic Influences   86
   - Unified Theory or Dimensional Shift   90
5. **SIGNS AND PROPHECIES**   93
   - Predictions   95
   - Dimensions   98
   - Orbs   103
   - Crop Circles   106
   - Ancient and Indigenous Prophecies   113
   - The Incas   120
   - Healers   125
   - Crystal Skulls   127
   - Apocalyptic Predictions   133
   - Modern Catastrophes and Predictions   137

6. CONSCIOUSNESS — 143

- *The Burgeoning Consciousness Movement* — 145
- *Consciousness Defined* — 151
- *The Oneness Movement* — 156
- *A Wealth of Encouragement* — 158
- *The Science of Consciousness* — 163
- *The Power of Intention* — 167
- *Science and Spirituality* — 175
- *The Universal Law of Attraction* — 180
- *Perennial Wisdom* — 186
- *Non-duality and Consciousness* — 189
- *A Common Vision* — 193
- *Our Energy Fields* — 196
- *Reincarnation* — 199
- *The Power of Thought* — 204
- *Sages through the Ages* — 209
- *Global Coherence and Mass Consciousness* — 212

7. CONCLUSIONS — 221

- *The Journey* — 223
- *The Web of Life* — 226
- *The Names of Source* — 229
- *Love and Heaven on Earth* — 243
- *How Do We Get There From Here?* — 248
- *Global Issues* — 264
- *The Perfection of Hope, Harmony and Love* — 275

Endnotes — 285

INTRODUCTION

The Journey

Would you like to experience a journey through space and time and beyond those dimensions? To the center of our Milky Way galaxy, out past the Pleiades and beyond what we know as the physical universe? We shall look from one side of this beautiful blue planet to the other, and then into the center of an atom. We will visit the dawn of time and the Mayans at the peak of their civilization; we will look at our present time and some possible futures. Other cultures, ideas, theories, and dimensions may help us stretch our imaginations and perspectives, so we can understand why 2013 is such an important beginning. I am going to share some information and views around this subject, but I am hoping you will come to your own unique perspective about these topics.

For you to understand why I believe that 2013 brings the very real possibility of greater peace on this planet, you may need to open your mind to some ideas you may not have encountered before or, if you have, perhaps you dismissed them as not within your realm of belief. We will even look at beliefs and see that they are just thoughts, and thoughts can change! They can open new worlds.

As you may have guessed, my research concerning this much-talked-about year of 2012 led to frightening prophecies, science vs. spirituality issues, and a thorough look at consciousness, for they all play a role in this inquiry. I was studying the Mayan civilization of Central America, planning to write a novel, and finding their culture fascinating. The Mayans had several ways of tracking time, but their Long Count Calendar really grabbed my attention; it seemed mysteriously prophetic the way it ended on December 21, 2012. I wondered, since I am an Aquarian by sun sign, how this lined up with the approaching Age of Aquarius. One book led to another. For years I had been reading about metaphysics which brought some of the latest research in quantum physics together with some of the oldest spiritual philosophies, revealing supporting ideas. Now some of these authors were writing about the Mayan calendar and how it was relevant to these inquiries.

There was quite a lot of "press" about this "end date." Some felt a cataclysmic "end of the world" was actually imminent. If one does a search on the Internet for "2012," an enormously long list emerges! One subject led to another as I read my way through at least sixty books on everything from astrology, astronomy, crop circles, archaeological discoveries, shamanism, mind-over-matter, and Stonehenge to the evolution of consciousness. I wondered what *was* in store for Homo sapiens at this time. My research was definitely not limited to scientific treatises, but included a broad spectrum of knowledge including channeled information. What did other ancient legends predict? What do scientists on the "leading edge" say? And most importantly, can we influence what our future holds? I found a lot that indicates we *can* have a huge part in creating how this world age ends and **if** we will have peace and harmony in the new era.

Astronomers (perhaps unknowingly) agreed with the Mayan shaman/seers when they announced that our Earth and sun would, indeed, cross the ecliptic (an imaginary line drawn horizontally through the 'thin' part of our spiral galaxy – think thin bagel) and line up with the "womb" (center) of our Milky Way galaxy around December 21, 2012. In 2003 scientists declared that this center of our galaxy is a "black hole" and the "birthing place of our world," as the Mayans had said. However, there is some disagreement about the black hole, especially from Dr. Paul LaViolette, a scientist of innumerable credits, interests and publications, who states that the center of our Milky Way galaxy or near it is a massive object that explodes periodically.[1] This sounds even more like a "birthing place." This correlation of modern science and myths from all over the world make this subject even more captivating.

Study on the subjects of astronomy and historical legends broadened and gave new meaning to one of my favorite quotes from an ancient source, Hermes: "as above, so below." The work of "Hermes Trismegistus" ('thrice-great Hermes'), *The Emerald Tablet* (circa 3000 BCE) is believed to have been born from a fusion of the sacred literature about the Greek god

Hermes and the Egyptian god *Thoth*, their great gods of wisdom; literature credited to 'Hermes' is probably the works of many great sages.[2] The two gods were worshiped as one in the Temple of Thoth in Khemnu, Egypt, which the Greeks called Hermopolis. The shortened quote above comes from the full quote, "That which is Above is like that which is Below and that which is Below is like that which is Above, to accomplish the Miracle of Unity" suggesting that "the experiences of our lives are reflections of events occurring on a much larger scale in the cosmos," writes Gregg Braden.[3] As well as the corollary that we affect the larger cosmos with our consciousness and lives, a concept we will explore in depth. Sharron Rose, filmmaker, writer, teacher, says, "We are the reflection of divinity, a fractal representation and mirror of the divine source."[4] Above - Below.

The ancient Vedic literature from India and the Cross of Hendaye in France show (in a mystical, alchemical code) that the Kali Yuga, or the Iron Age, will also end around 2012. So we will examine these as we 'journey' on. I dug deeper, expanding my search from *Earth in Upheaval* by Immanuel Velikovsky,[5] to *The Divine Matrix* and *Fractal Time* by Greg Braden,[6] to books by Solara and Barbara Hand Clow, all giving different perspectives. Yet, most were confirming: "As in the macrocosm, so in the microcosm."

I believe the Mayans never intended to predict an apocalypse or cataclysmic end of the world, instead, just a notice that a World Age would be ending, a history of the catastrophes that had occurred at the end of other world ages, and a message that consciousness controls what that transition would be like. In these pages I hope to reveal just how powerful the human mind/consciousness is, and that we do have time to direct positive, caring, healing thoughts and feelings toward our planet, our leaders, and our fellow humans. We can bring in the next age in a peaceful manner and confirm that *"All truly is well."*

I explored many conflicting ideas, not all scientific (and even the scientists routinely disagree!) I hope to be of value, not to a dissertation committee, but to those who have heard the "buzz" and just don't know what to believe! If you can only consider the "accepted" scientific viewpoint, you need go no further; we will look at some perspectives that are far from that! I frequently find fascinating ideas and information 'out there in left field!' I will present some theories that even I (an Aquarian intrigued with science as well as spirit) have found challenging, to see if we can get a good sense of why we need to examine 2013. The goal is to find the higher truth, a fusion, perhaps, of the pragmatic view and the more mystical outlook. There have to be people who are willing to hear and examine novel ideas for thought to evolve and move forward and, 'Thank goodness,' there have always been some – a minority – who will! (Remember Columbus?) One does not have to rely solely on the scientific view or the mystical/spiritual; they can come together and add much to our understanding. In fact, an end to 'either/or' thinking and duality is one concept we will investigate in the area of consciousness. If you have been following the 'consciousness' movement awhile, some of this will be like a review; for others not yet 'into' that scene, we will build a foundation and go on to recent developments in an arena that is changing so rapidly it is hard to keep up with!

In reading *Ancient Mysteries*,[7] by Peter James and Nick Thorpe, rather late in my research, I was confronted with the scientific 'party line' debunking a few of the other 50 or 60 books I had read! Should I throw out my original premise? I found they quoted only the part of the science that seemed to disprove an ancient mystery in many cases. For example, a primitive view such as "God removed two stars from the Pleiades and allowed the water from the two holes to flood the earth" (to explain that a comet or meteorite coming from the direction of the Pleiades constellation may have caused the floods) might seem like superstition and give a scientist pause. However, I found that many of the ancient legends did have a strong historical basis, though explained in terms unfamiliar to our way of speaking. James and Thorpe totally dismiss Atlantis, but do concede that there have been many geomagnetic reversals, and many catastrophic events, at least one of which may have caused crustal displacement. This has been made very clear by the discovery of tropical plants in the mouths of extinct animals in gigantic heaps within the Arctic Circle! Could crustal displacement have moved Antarctica to its present position? Let's look at the evidence: equatorial flora and fauna were pushed into what is now the Arctic Circle, the last Ice Age was brought to an end rather suddenly, and, let's not forget – at one time the continent of Antarctica had tree ferns once common to sub-tropical Africa, South America and Australia (deduced from ice core samples).[8] We will find on our 'journey' that our Mother Earth has been a very active "entity" for a very long time.

It is well known that there have been numerous "cataclysmic" events on this globe. Not that it is catastrophic, but even now her magnetic North Pole moves about 40 meters a day, usually northward. It has shifted, in historical times, from near Hudson Bay to northern Canada, near an island north of Resolute Bay at present, moving toward Siberia. Since this rate of movement has been increasing lately, Fox News[9] reported on 1/6/11 that the Tampa International Airport was forced to readjust the input regarding its runways (re-designation of numbering on aviation charts) to account for the change in the Earth's magnetic North pole, which pilot's rely on, of course, to navigate. St. Paul, MN airport did the same several years ago.[10]

What is happening with the Earth and what the ancient prophecies have told us begin to seem complicated, as well as comparable. So we will examine some archaeology, history, geology, mythology, astronomy and astrology (Yes, far out!) and more, to set the scene for looking at the world through the eyes of the Mayan people. So 'hang' with me! It will be an exciting 'journey.'

I use the abbreviations *BCE* (meaning *before the current era, Before Christ,* or *before 0 AD)* and *AD* (for our current era) throughout because it seems clearer than the recent change to CE (current era) used in some literature. I also use gender terms as they are quoted, which sometimes appears sexist, but *I* usually speak in general terms of human or either sex that seems appropriate. Where I have emphasized a word or phrase in another author's work with italics or have added an explanation in parentheses, I have placed an asterisk * rather than making a note each time.

THE MAYANS

The Ancestors

The Mayans started all this "twentytwelveology," as John Major Jenkins[1] calls the "buzz" that is going around about the end of the Mayan calendar. There are many versions of and perspectives about this phenomenon, but it seems to have all started with the people from whom the Mayans descended – the "mother culture" – the Olmecs of eastern Mexico. However, there are correlations and corroborations from all over the planet about this significant year.

Did the ancestors of the mysterious Olmecs come across the Bering Strait 13,000 years ago at the end of the last Ice Age or more than 40,000 years ago, or were they "born out of a cave in the Underworld" as one legend relates? Mitochondrial DNA studies suggest that some Native Americans may have diverged genetically from the Siberians as early as 20,000 years ago.[2] Modern scientific dating methods seem to be pushing all previous settlement dates back. Worked stone tools and charcoal found at Topper in South Carolina by Dr. Al Goodyear (University of South Carolina) have been radiocarbon dated to around 50,000 years ago (some dispute by other archaeologists); evidence of humans at the site is dated to 20,900 BCE.[3] Stone tools, wooden implements, and hearths, found near Monte Verde, toward the southern tip of Chile, have recently been dated to around 22,000 BCE (other reports say 33,000 BCE).

One of the early well-established cultures of North America, the Clovis culture, is dated as early as 13,500 BCE. Recent archaeological finds at the Gault dig (a Clovis site) in Texas are exposing occupation spanning several hundred years and deep tests show evidence of an even earlier occupation.[4] These people were primarily hunters of megafauna: extinct animals such as the mammoth, bison, horse and camel, which all disappeared from the

North American continent some 11,000 to 12,000 years ago. It is curious that about 70% of the Clovis sites in North America have what has been called "an organic black mat" over the top. This "peaty covering" coincides with the extinction of mastodons, sloths, dire wolves, tapirs, palaeolamas and short-faced bears in addition to those animals mentioned above. This indicates an extended period of wet and cloudy weather. Just beneath the mat in all sites lies a layer of magnetic grains with iridium, glass-like carbon containing nanodiamonds, and fullerenes with extra-terrestrial helium. These represent the detritus of an explosive object – a comet, or some say an asteroid – which destabilized the Laurentide Ice Sheet that covered much of North America, causing fires to the south and rapid melting of the ice.

At the 2007 meeting of the American Geophysical Union one session explained that the "black mat" followed the explosion of a comet which was postulated to have hit near the Great Lakes or broken into pieces over the enormous ice sheet that covered Canada and the northern part of the U.S. at that time.[5] (We shall examine other localized and planetary catastrophes from this time period later in our 'journey.') Did the catastrophic event or the extended lack of sunlight or both cause the extinctions and a change in the culture? Were the ancestors of the Olmecs from the Clovis culture? Most scientists say "no," though there is some evidence of that type culture in northern Mexico. The cultural markers are different further south. It may be that this cataclysm (around 10,000 BCE) is the one spoken of by the Mayans as the end of their Second Sun (we are now almost to the end their Fourth Sun or World Age which began 3114 BCE). On the other side of the Atlantic at this same time (10,000 BCE) the Rig Veda was composed in India, the Sphinx in Egypt was built (some disagreement on this date), metal blades were being used in France for reaping barley, and the Sahara was a lush, green savannah.

Were the Americas peopled from early Celtic Europe, southern Asia and the Polynesian islands, or the Mediterranean area? Others give credit to the early Phoenician sailors. Boats of tortora rush (reeds) almost identical to those in which the Pharaohs sailed the Nile have been crafted near Lake Titicaca, Peru, for hundreds, perhaps thousands of years and are still made at Suriqui, Peru. Scholars have confirmed the likeness, and legends there say the knowledge of these boats was given to them by the Viracocha people when Tiahuanaco and the surrounding area were much wetter. Geologists now say nearby Lake Titicaca, 12,500 feet above sea level, and the whole Altiplano is *still* gradually rising, especially in the northern part. The strand line on the lake is tilted, so we can deduce that this part of the hemisphere also suffered catastrophe at some point. So how did these "teachers" and ancestors of the Inca people come to Peru? Some say they travelled all the way from the homelands of the Olmec on the Gulf coast. Or were the Olmecs related to the ancient culture discovered by archaeologists off the coast of Brazil? It is not known, but few scholars believe the Americas were peopled from Polynesia.

Some posit that seafarers from the Canary Islands were blown westward to the Caribbean. Legends similar to the Mayas' have turned up in West Africa. The giant head sculptures originally at La Venta (an early Olmec city), some weighing as much as 30 tons, have African features, but some of the figurines found there appear almost oriental.[6] Several Mesoamerican myths describe bearded men (full-blooded Amerindians have very little beard)[7] who came from an "island" to the east and brought a philosophy of peace, and knowledge of astronomy, mathematics, and pyramid architecture. Atlantis? Yes, we will look at that point of view! (Who can argue with Plato?) Others say the Mayans' knowledge of the calendar and its vast timeline was gained from visionary "journeys" with hallucinogenic substances. Certainly, it is hard to imagine how they could have known that the center of the Milky Way was the "birthplace" of our world in any other way! From all that I researched, the theory of "multiple settling" of the western hemisphere is the most plausible. Sites in the eastern U.S. and cave paintings and hearths from the "white forest" in Brazil that have been carbon-[4] dated to 36,000 and 17,000 years ago, respectively, help us come to this conclusion.

From a few ancient oral legends, some stone effigies, carved stelae (rock columns), and a few ancient step pyramids, now mostly covered by the jungle, many scholars place the beginning of the Olmec culture around 2000 BCE. This is contemporary with the building of the huge timber and stone monuments at Stonehenge and the destruction of Sodom and Gomorrah.[8] (Interesting side note: apparently Bab edh-Dhra (aka S & G) sat on a fault line, and was destroyed by an earthquake that forced combustible material to the surface; the ash found in the archaeological dig indicates that the cities burned after they were leveled.)

As with many ancient cultures, not a lot is known about the Olmecs, so they seem mysterious. The Books of Chilam Balam (Jaguar Priests), ancient Mayan religious texts, reported that the first inhabitants of the Yucatan were the "People of the Serpent."[9] Their legends, which variously call the leader Itzamana, Kukulcan, or Quetzalcoatl, say he could cure by 'laying on hands' and that he "revived the dead."[10] They arrived from the east in "a raft of serpents."[11] Carlos Barrios, a modern shaman and member of the Mayan Elders Council, reports that in ancient stories told in the most traditional tribes, the Maya descended from a culture/place called Tula or Tulán, which also gave rise to cultures in Africa, India and Tibet. This original civilization, they say, was located on an island continent, now gone, in the Atlantic Ocean. (Note the name of the ocean from which they arrived.) The *Popol Vuh*, one of the few surviving Mayan stories, relates, "From there they came, from the East. . ." The *Annals of the Kaqchicheles* tells a similar story. The Mam (a Mayan tribe) Grandfathers point to the ruins at Tulum as the only one facing the ocean and as one of the oldest. Some of the oldest settlements are found in Mam territory in western Guatemala.[12] A pyramidal temple found in the Hacavitz Mountains of Guatemala, is claimed by the present-day Mayans to be 10,000 years old, which places it before the Olmec culture, and possibly contemporary with at least one of the early pyramids in Egypt. Barrios contends that, from the oral traditions and stele with dates thousands of years in the past, the Mam tribe existed before the catastrophe. But which catastrophe? Was it the one which nearly wiped out the earth in the 11th millennium BCE or a lesser catastrophe at the end of the Third Sun, 3114 BCE? This date may be the one of which the Judeo-Christian scriptures speak, from which Noah (of Old Testament legend) survived. Or perhaps the Mayan ancestors survived both! (Not the same persons, obviously!)

We call the time before written history, before 9,000 BCE, *pre-history* (though some scholars place the writing of the Rig Veda of India prior to that).[13] We tend to assume that humankind prior to that date was primitive cave dwellers, but that stage of Homo sapiens occurred long before the era we refer to as The Blank. Although it appears that humans sprang "full blown" into "civilized" societies about 7,000 to 8,000 years ago, this was not the case. I believe, along with Barbara Hand Clow, Hunbatz Men (a present-day Yukatek Mayan spiritual guide from the Itza tradition), authors Gilbert and Cotterell [14] and many others that there were highly developed cultures prior to 10,000 BCE. There were cities, governing councils, ocean-exploring navigators and a degree of technology. What happened to cause this "Blank" in the timeline and the almost complete erasure of previous accomplishments? We have alluded to and must now examine a worldwide catastrophe. The ancient texts speak of disasters that changed the planet. Mayan history tells of previous "suns" (ages) that ended cataclysmically. The geology confirms this.

Catastrophes

Ancient reliefs in the Osireion Temple at Abydos depict the Pharaoh holding a pillar tilting 20 to 25 degrees off vertical (same as the axial tilt of our planet), then placing it upright to reestablish *Maat* (order on Earth). The four sections on top of the Djed pillar (See Illustration #1) are considered symbols for the equinoxes and solstices.[15] Does this illustrate the event that tilted our planet?)

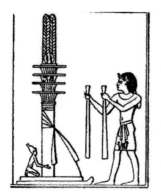 

Djed pillar, Abydos, Egypt

Such an occurrence would have set back all civilizations on earth at the time. Such mass confusion and devastation could easily have led to The Blank in historical records. Can you imagine what would happen if a major catastrophe of that scale happened today?

Archaeoastronomy has proven that some constructions at ancient sites in Egypt, the Americas, and other places were aligned with certain stars on specific dates and these dates have often been verified by radiocarbon dating.[16] With each passing year scientists are uncovering more evidence of past worldwide catastrophes. There is much more to our delicate relationship with our near and far stellar and planetary neighbors than is widely discussed. Astronomers, with the help of satellite telescopes, are becoming much more precise in their calculations of sunspots (cycles of intense magnetic storms on the sun), electromagnetic radiation from the center of our galaxy, our solar system's path as it circles around our Milky Way galaxy, quasars and how all of these affect the residents of planet Earth.

We can now put all that together with the findings of geologists and anthropologists to form a picture of what may have happened in the past. Many areas north of 55⁰ latitude (near the Arctic Circle) have great heaps, up to hundreds of feet thick and miles long, of extinct and modern animals thrown together, some with green vegetation in their mouths, bones broken, unputrified, as though frozen suddenly, along with masses of broken trees pushed together with sand and gravels. Vegetation remains found in these masses within the Arctic Circle is tropical in nature, indicating it grew nearer the equator. It seems all too massive, dislocated too far to be explained by a tsunami and too sudden to be explained by an ice age. The only reasonable explanation appears to be a shifting of the Earth's crust (lithosphere), perhaps combined with tidal waves that could cause such a huge worldwide disaster.

As with the evidence of the demise of the Clovis culture, the ash deposited at many sites indicates volcanism accompanied or followed the first wave of destruction. Scientists now say the Earth occasionally passes through areas of large clouds of interstellar dust which would carry an electromagnetic charge that would have a *thermal* effect on the planet, but this would probably not have had a sudden consequence. It could, however, account for the advance and retreat of the Ice Ages prior to 11,000 BCE. One or several meteorites could cause great damage and set off volcanoes depending on their size, but a large comet or planetoid hitting or coming very near the earth could certainly pull the nearest point of magnetism to another location, since the Earth's crust (only 3-25 miles thick) sits on an 1800-mile deep layer of molten magma.

The Yellowstone underground caldera, one of the largest known in the world, last erupted about 640,000 years ago, so geologists say. Whether this was precipitated by extra-terrestrial impact is not known, but it covered the North American continent in several feet of ash/dust and probably brought about a huge climate change. The comet or asteroid that hit near the Yucatan peninsula 65 million years ago, making the 175-kilometer Chicxulub (pronounced cheek'-sha-loob) crater, was the event that wiped out the dinosaurs and a large percentage of other species. 26 to 28 million years ago saw the last eruption of the La Garita in southern Colorado, leaving a huge caldera, one of the larger in the world. In times past this globe has taken some serious "hits."

These cataclysms bring up a more recent disaster: the one that Plato affirmed destroyed the island nation of Atlantis completely. So, why do we want to attribute this input of knowledge into the western hemisphere to the survivors of Atlantis? First, there are common elements, such as ziggurats/pyramids, legends/myths, and anthropological and archaeological discoveries that are similar from Egypt, Tibet, central and South America, even Asia, that appear to have a common source.

There are step pyramids and ziggurats (temple pyramids with larger terrace levels) distributed all over the globe: Indonesia, Cambodia, Sri Lanka, Europe, China, and Africa (different types), as well as Mexico and Central America.[17] The Great Pyramid of Cholula (known as *Tlachihualtepetl*, which is Nahuatle or Mayan for "artificial mountain"), a four-tiered ziggurat, is the world's largest pyramid by volume (according to *Guinness Book of Records*) with a total volume estimated at over 4.45 million cubic meters; it is 1476 x 1476 feet at the base. (The Great Pyramid of Giza, Egypt, is about 2.5 million cubic meters.)[18] Only one side of Cholula has been uncovered, but five miles of tunnels have been excavated, proving it was built over many centuries, by possibly four cultures, and may be as much as 3,000 years old. A mural discovered at Cholula, called the Drunkards, probably portrays the effects of a potion made of peyote. A circular step pyramid in the same area near Mexico City, partially excavated in the 1920s, had been buried on three sides by lava from a nearby volcanic eruption. An archaeologist for the National Geographic Society dated the ruins to around 6500 BCE due to the lava.[19] It appears the building of pyramid-like structures started rather suddenly in the western hemisphere.

So why did the ancestors build pyramids and pass that knowledge on to the Mayans? It may have begun when the early people venerated the volcanoes as the source of creative force in their world.[20] The volcano now called San Martin Pajapán probably loomed large in the lives of the people living around the Veracruz Gulf coast, having witnessed "the world being born out of the Underworld below."[21] They built a symbol of it in the town of La Venta in the form of a huge (25,000 square yard base) earthen pyramid with fluted sides. In front of it the builders laid layer upon layer of serpentine blocks (a rock of magnesium silicate, usually green or mottled in color) in patterns symbolizing their connection to the spirit world and supernatural beings. Others believe this architectural expertise was taught by masters from one of the early cultures that did not survive the catastrophes. The aligning of their cities to the cardinal directions (NSEW) and the centering of buildings and streets precisely according to the shaman/ruler's interpretation of the cosmos and stars at that time was divined in visionary journeys. Their pyramids resembled the sacred mountains and were seen as portals to the Otherworld. They built their myths and Creation stories around what they saw on earth and in the night sky. They truly believed, "as above, so below." Were the ancestors taught a new technology by the "gods" that arrived from "the east?" The hieroglyphics and engravings found in Olmec ruins show without a doubt that much of their cosmology was passed on to the Mayans. So what was this "island to the East" that keeps coming up in the legends? There are no islands east of the Caribbean until you get to The Azores or the Canary Islands.

Atlantis

Why do so many writers say one of these two groups of islands is all that remains of Atlantis? Firstly, the timing agrees with a known disaster around the tenth millennium BCE, the earliest description of which that we have was written around 360 BCE by Plato. Plato, whose real name was Atistocles, (born in 427 BCE) was initiated into The Mysteries at age 49 inside the Great Pyramid. He received instruction from the wisest of the Egyptians and the Chaldeans for thirteen years. There is no doubt that he was overly qualified to speak on almost any subject. Yet he was criticized during his lifetime for revealing many of the advanced ideas and secret principles that he learned in Egypt.

Plato received his information about Atlantis from Critias (circa 460-403 BCE). Critias learned of Atlantis from stories passed down to him by his great-grandfather, Dropidas, who had the original writings (translated into Greek) of Solon, the Athenian statesman (639-559 BCE). Solon had stayed in Sais, Egypt for ten years and, in talking with a priest of great age, discovered that the Greek myth of Phaethon, was actually based on facts from nine thousand years previous. The priest had said, "Now, this has the form of a myth, but really signifies a declination of the bodies moving around the earth and in the heavens, and a great conflagration of things on the earth. . ."[22] The scene portraying the Djed pillar in the Osireion described above may have been the Egyptians' way of describing what had happened to the earth in their ancient history. Critias, in the dialogue, said that his descriptions of Atlantis were precisely in the words of Solon. The aging priest had told Solon of how the men of Atlantis had attacked areas within the Mediterranean Sea, but had been pushed back and defeated by the Greeks. "Afterwards there occurred violent earthquakes and floods;

and in a single day and night all . . . sank into the earth and the island of Atlantis in like manner disappeared in the depths of the sea."

At the beginning of this discourse written by Plato, Socrates had asked, "And what is this famous action of the Athenians, which Critias declared, on the authority of Solon, to be not a mere legend, but an actual fact?" Would Plato have disrespected his grandfather and one of Athen's greatest statesmen, to assure that it was fact when it was a fiction? His chronology would place the invasion by Atlantis into the Mediterranean area around 10,000 BCE and, apparently, its destruction shortly thereafter.

Crantor the Philosopher (3rd century BCE), who wrote the first commentaries on Plato's dialogues, went to Egypt and was told that the column of hieroglyphs that Solon saw there describing the event was still preserved as of 300 BCE.[23] We will note that Edgar Cayce (the famous Sleeping Prophet), as well as others, spoke without any doubt about the existence of Atlantis. (I warned you we would pull information from many sources!)

Abbé Brasseur de Bourbourg in the late 1800s travelled widely in Mesoamerica, learned the Mayan dialects and hieroglyphs, translated the *Popol Vuh* and the *Madrid Codex* (two of the few surviving Mayan documents), and found surprising similarity between the Mayan and ancient Egyptian alphabets as well as other parallels of culture. The serpent was an Atlantean emblem of divine wisdom. This is comparable to the "plumed serpent" title applied to Quetzalcoatl or Kukulcan, the wisdom-bringer of ancient Mayan legend. Plato described in detail the "City of the Golden Gates" with its great pyramidal temple in the center. Many believe this architecture was brought to both Egypt and Central America by survivors of the calamity that eradicated Atlantis. Others point to the purity of the Ayamara language still used in the highlands of Peru and find it so similar to that of the Basques of northern Spain, postulating that both may be from Atlantis as it is not found elsewhere.

We saw in the discussion about the Clovis culture (c. 12,000 to 10,500 BCE) that it ended when some disaster, possibly asteroid chunks, hit the Laurentide Ice Sheet of North America, causing fires and rapid melting of the ice around 10,500 BCE. Whitley Strieber writes of "a sudden dump of freshwater into the North Atlantic from a gigantic glacial lake in Canada" around this same time.[24] An Aztec legend relates that hot volcanic dust and rocks rained down from the sky followed by great floods in the Valley of Mexico. The Algonquins, Iroquois, Chickasaws and Sioux have similar flood legends. Stories from indigenous cultures worldwide (the Tupinamba Indians of Brazil, Araucanians of Chile, the Pehuenche of Tierra del Fuego, Inuit of Alaska) speak of "the day the world shook," darkened skies (a dim sun), and bituminous, burning rain. There is evidence of devastating floods in the Nile Valley during this time period. Geologists, archaeologists, and pre-historians have confirmed that the eleventh millennium BCE was important in Egyptian prehistory due to prolonged natural disasters that reached a crescendo just after 10,500 BCE.[25]

Hundreds of miles of ice fields with mountainous piles of jumbled, broken Pleistocene (now mostly extinct) animal carcasses, trees, rocks, and gravel which had been flash-frozen were found in northern Canada, Alaska, Siberia, and Russia. This would take a tidal wave of enormous proportions! Paleontologists report that around 10,000 BCE, well over 200 entire species (millions of individual animals) disappeared. Most of the remains found within the Arctic Circle were adapted only to a temperate climate. Andrew Collins relates in his *Gateway to Atlantis* that the *Chilam Balam* of Chumayel, "spoke of an almighty cataclysm during which the "Great Serpent" was ravished from the heavens so that its skin and pieces of its bones fell upon the Earth."[26] The Calina Carib tribe of Suriname, South America, retains myths of a fiery serpent that came from the Pleiades (pronounced plee'-a-deez, one of the nearest star clusters to the Earth) and "brought (their) world to an end" with "great fire and a deluge."[27] Garcilaso de la Vega, chronicler of the Incas (16th century AD), confirmed from Incan legends that after the waters of the deluge had subsided (legends say it covered the mountains), there appeared in the country of Tiahuanaco a man of miracles called Viracocha. This man seems very similar to Kukulcan (or Quetzalcoatl) who, according to the legends, appeared from "the East" to teach and help the ancestors of the Olmecs.

I hesitate to quote Immanuel Velikovsky due to what I think are errors in his dating, but his descriptions of the effects of worldwide catastrophes that have occurred are detailed and copious. In *Earth in Upheaval*[28] he describes evidence that the Himalayas are young, relatively speaking, having risen to their present height within the last Ice Age (which ended around 10,500 BCE, same time as our "catastrophe"). The east African valley called The Rift (the longest meridional land valley on earth) appears to have also fallen under the power of whatever hit the Earth with enough force to cause crustal upheaval and buckling. The bare fault scarps and the fact that the natives there have traditions of great changes in the structure of the country led Velikovsky to agree with geologist R.F. Flint that this happened near the end of the last ice age. He refers to a book from 1927 which describes fields in Arabia called *harras*, some hundreds of miles in diameter, of millions of sharp stones that appear to be from a "train" of meteorites. In some areas there, previously habitable, water levels dropped and rivers ceased to flow around this time.

Velikovsky also quotes from the *Bulletin of the Geological Society of America*, 1952, concerning the more than 500,000 oval craters of various sizes scattered over the eastern seaboard of the U.S., commonly called the Carolina Bays (named for the bay trees in the area). The consensus in the early 1900s was that they were meteoric, but that has been challenged by the fact that few meteoric materials have been found in the depressions. The native people of the area called the depressions *pocosins*. The best explanation for these impressions, not clarified by Velikovsky, is a large comet, since comets are made of ice and frozen gas. If the main impact had been near the eastern side of The Bermuda Triangle, it seems fragments of ice could have covered much of the southeastern coast. The Puerto Rico Trench near the southeastern corner of the Triangle is over 30,000 feet deep, one of the deepest areas in the Atlantic. Andrew Collins in his book, *Gateway to Atlantis,* proposed that a comet hitting in this area could have resulted in the tidal wave that wiped out Atlantis.[29] He believed that Cuba is but a remnant of what was there earlier. Some have placed this cataclysmic occurrence around 10,000 BCE. A familiar date, we are finding. Did they suspect it was coming? The only thing left directly west of the Straits of Gibraltar (called the Pillars of Hercules by Plato) are the Azores. Could they be the mountain tops that were described on the island of Atlantis? In that region of the Atlantic there is a long finger of the Atlantic Ridge that is only 200 meters beneath the surface in places. Erratic boulders (formed by glaciations) on the Azores indicate that it was either once much closer to one of the poles or that they were placed there by an enormous tidal wave.[30] Also, it appears that limestone on the tops of some summits of the Atlantic Ridge may have been above water within the past 13,000 years. Think how much lower the oceans were before the rapid melting of the Ice Age.

If a comet precipitated this cataclysm, do present-day astronomers know of it? Scientists of all the pertinent categories cannot even pinpoint (or agree upon) the cause of the sulphuric acid content in ice cores from 532 AD! They think this acid content might have been laid down by an earlier eruption of the Krakatoa volcano that split the island of Java forming Sumatra. Today there are astronomers and satellite telescopes watching the skies and feel they can predict if a comet, asteroid or meteorite (one weighing 60 tons was found in Namibia) comes near the Earth. We tend to forget that we live in a "busy neighborhood" with interplanetary dust clouds, gamma rays from exploding supernovas, an elliptical (non-circular) orbit that takes us closer then farther from the sun, comets with unstable orbits, and an Asteroid Belt between Mars and Jupiter, just to name a few! We see the effects of some of these anomalies worldwide: Barringer Crater near Winslow, AZ and the destruction near Tungusta, Siberia where a meteoroid 120 feet across, weighing 220 million pounds exploded in the atmosphere after heating to 44,000 degrees F. These are just two of thousands of clashes with 'foreign' objects on our "journey" around the galaxy.

Barbara Hand Clow agrees with Allan and Delair (*Cataclysm*)[31] in her book, *Catastrophobia*,[32] in placing the cause of the disaster around 10,500 BCE as the Vela supernova explosion (she calls the remnant that hit earth Phaeton).[33] She dates the disaster at 9500 BCE; I suspect from all that I've read that there were multiple events over several hundred years, especially the worldwide flooding as previous ice caps rapidly melted. A supernova is a massive star that has used up all its fuel and in its final collapse creates a massive explosion that propels it into space. The Vela Supernova Remnant (still an enormous expanding debris cloud) is 815 light years away; so what is the problem? In their collapse supernovae, in a period of about one second, set off thermonuclear explosions that are hot enough to fuse carbon and oxygen into heavier elements sending out incredible shock waves of gas and dust. These waves contain gamma rays from the decay of titanium-44 that form nitrates in our upper atmosphere which deplete our ozone layer. These nitrates after falling to earth are found in ice core samples that tell us of past events of this type of exposure. Scientists estimate that a Type II supernova explosion would have to be closer than 26 light years to destroy half of our ozone layer. Light from the Vela Supernova reached the earth about 11,000 - 14,000 years ago, but were the electromagnetic waves blasted off strong enough to cause the kind of damage of which we have evidence? Fragments theoretically could have arrived a few hundred years later.

NASA reported on a meteorite in 1996 that they said hit Antarctica about 13,000 years ago. (Familiar date? 11,000 BCE) It is now suspected that the 2004 tsunami in Indonesia caused by a 9.1 Richter earthquake may have a connection to a gamma ray blast (100 times more intense than any previously recorded discharge) that was measured on Earth 44 hours after the quake. It was determined the burst originated from a neutron star located 20,000 to 32,000 light years from us, or about the same distance as our galactic center.[34] What type of magnetic disturbance preceded the radiation? Was this type of occurrence sufficient to cause crustal displacement? Probably not that alone; but something really big influenced legends, as well as the structure of the earth worldwide at the end of the last Ice Age.

Galactic core superwaves of ionized gas are postulated to cause major solar storms which send out even more electromagnetic waves toward Earth. Some think there was an event of this nature around 10,700 BCE. We will go into this in more detail when we look at the Mayan calendar and Astronomy. In the last 6000 years these blasts have occurred on an average of about 500 to 700 years between blasts, but can be as much as 2000 years apart; we are now 700 years without one. Do these blasts "perturb" (as they say in scientific circles) the orbits of loose comets or asteroids? Has this happened in the past? Let's look at the legends.

Myths, like archetypes, have value in many ways. In the case of prehistory, they stand in for history and in many cases have proven to be quite accurate, though cloaked in symbols. The Stones of Ica found in Chile are believed by some to date to a period prior to these catastrophes. They graphically depict people using telescopes to view an object that looks like a comet. In the same area, the Incan legends, recorded by a priest, Father Molina, spoke of the ancestors of the Incan people who perished as "the waters rose above the highest mountain peaks" and the wind carried a man and a woman in a box to Tiahuanaco. In another account, by Garcilaso de la Vega, "After the waters of the deluge had subsided a certain man appeared" . . . Viracocha.[35]

In South Africa the Zulu people speak of *Mu-sho-sho-no-no*, a star with a long tail, that came to Earth thousands of years ago turning it upside down so that the sun rose in the south and set in the north. (Does this sound like a possible crustal and/or axial displacement?) Their shamans predict it will return in their year of the "red bull," which is 2012. It is hard to understand ancient legends without being aware of the people's closeness and connection to the Earth and the cosmos in which it travels. "As above, so below" was a very strong principle as they considered the small and large cycles of which they were a part. The Maori of New Zealand have a ceremony which they say is performed every 13,000 years (previous one approximately 11,000 BCE, a familiar millennium) which is one-half the precessional cycle of 26,000 years. (We will address this in later chapters.) Drunvalo Melchizadek describes being invited to take part in this ceremony a few years ago at a large, square "Celtic-type cross" cut a foot deep into the ground called the "center of the world." The legend states that "people" from Sirius (another recurring constellation theme) placed a huge crystal under this spot. They honor the cetaceans from whom, they say, mankind evolved, and share this belief with the ancient Sumerians.[36]

Back to the Olmecs and Mayans: the Troano Document (an ancient stone found in the Yucatan) was translated by Augustus Le Plongeon in the late 1800s as describing the sinking of Atlantis.[37] Many Mayanists disagree with his interpretations but his photographs of many Mayan ruins have been invaluable. Mayan glyphs on stone found in the astronomical observatory at Copan are adjacent to an engraved mammoth, which we know disappeared with the Ice Age. Also, a mural in Cerro el Hacha, Costa Rica, portrays a mammoth surrounded by Mayan glyphs.[38] Some feel these indicate an advanced civilization in Central America prior to 10,500 BCE.

Melchizadek explains that the secret Nakkal Priesthood knew they would have to leave Atlantis and take the powerful teachings of world ages to the places that the "Serpent of Light" (a spiritual focus point with a planetary physical effect) would eventually settle. The inner priesthood travelled to Tibet and the outer priesthood to Egypt and to the Yucatan eventually. There have been rumors from early times that a Great White

Pyramid existed near Tibet and a pilot in WW II said he took a photo of it. Melchizadek spoke with a team of explorers who had climbed high into the mostly inaccessible western Himalayas in the 1980s and actually found the pyramid! This is, he believes, to be a relocation point of that early priesthood. The pyramids that are found in China have been covered, over the expanse of time, with vegetation, but there are no reports of the Great White in recent years.

In reading about this esoteric group, I wondered if this priesthood was aware in some way of the relationship of consciousness and the magnetic balancing mechanism of Mother Earth. This brings up the question of whether focused prayers/thoughts can affect planetary conditions. Elders of the modern Mayas have cooperated in prayer sessions over the past few decades with other indigenous people of this hemisphere to bring the Serpent of Light (a spiritual focus point) to its new dwelling place in the Andes Mountains of South America. A prophecy by the Guatemalan Mayas had predicted that once it was in place a time of peace and spiritual growth would begin on February 19, 2013.[39] (I particularly liked this date since it follows my birthday on the 18th!) Carlos Barrios writes that after a cleansing period of chaos, *Oxlajuj Tikú*, the period of light, harmony and love will begin in 2014 – the beginning of a slow progression.[40]

The ruins of Tula (or Tollán), Mexico, had a low terraced and stepped pyramid/temple crowned with human figure-type pillars, called 'Los Atlantes,' so it is known as the Temple of Atlantes. The walls of the base are covered with slabs of volcanic tuff.[41] Behind the temple was a wall known as *coatapantli*, or serpent wall, which had apparently at one time surrounded the pyramid. It is easy to relate these serpents to Quetzalcoatl (the Plumed Serpent) and the name Atlantes to the island of Atlantis.[42] This was one of the main population centers of the Toltecs in 980 AD and is estimated to have had 30,000 to 50,000 people. U.S. Congressman Ignatius Donnelly (1863-1868) researched and compiled an impressive catalogue of the similarities between the myths, anthropology, linguistics and religion, of cultures in Mexico and Central America and Atlantis in his book, *Atlantis, the Ante-diluvian World*.[43] He is known for his early theories of catastrophism related to an impact event affecting ancient civilizations.

Hunbatz Men, the above mentioned Mayan elder of the Itzá tradition, in a talk given in 1998, spoke of his (ancestral) "people" from "Atlantiha" and how they understood the cosmic wisdom and spiritual work. He told of how his ancestors were told they must emigrate and teach new initiates the sacred symbols of the inherited wisdom, those symbols of the snake and the eagle which were related to the development of consciousness and the powers of the chakras. "With their sacred language they created the word *Hunab K'u*, . . . so they would represent the great concept of the creation of the universe."[44] This reminded me of what Edgar Cayce had said, "Iltar, of Atlantis, with a group of followers of the house of Atlan, the followers of the worship of the ONE . . . came westward, entering what would now be . . . Yucatan. And there began, with the people there, the development of a civilization that rose much in the same manner as . . . the Atlantean land. . ."[45] References to Atlantis come from many places.

One fact that seems to prove a link between Egypt and the western hemisphere is that traces of tobacco and cocaine, two plants that grew exclusively in the Americas, have been found by chemical analysis in the ancient mummies studied by forensic archaeologists at the University of Manchester, England and the Institute for Anthropology and Human Genetics in Munich.[46] Plus, Thor Heyerdahl proved that crossing the Atlantic in a small sail craft was quite possible.

Velikovsky reported on an Atlantic expedition that found beach sand in the middle of the ocean (where one usually finds fine sediment or 'muck'); beach sand is only found near continental shelves and coastal rims. This was 1200 miles from a coast and three miles down! Another interesting discovery there was granite and sedimentary rocks marked with deep scratches or striations, like those found where glaciers have moved over the land.[47] If I have not convinced you of the existence of Atlantis, consider the ancient maps that show a continent between Africa/Europe and South and North America. Is it possible this was the continent of Antarctica before it was pushed southward with the crustal displacement of 11th millennium BCE or a previous catastrophe such as the one that created the Chicxulub (pron. CHEEK' sha loob) Crater off the north end of the Yucatan 65 million years ago that eliminated the dinosaurs? The island of Atlantis, I believe, was smaller than many authors postulate.

There just seems to be too much evidence to rule it out the fact that there was an advanced culture that dispersed and had an influence on the Olmec culture and therefore the Mayans. There appears to be no doubt that there was a massive cataclysm approximately 12,000 to 13,000 years ago, and, regardless of the cause, that it was gigantic enough to move parts of whatever had been near the equator to inside the Arctic Circle and flash-freeze them! Since the Mayan calendar has come under investigation many researchers have been trying to pinpoint the exact conditions of that time on the Earth and in the "sky," especially since so many other scriptures and legends say it will happen again in 2012. Our tendency to want to know what is in our future is very strong!

The Olmecs

Though much of the Clovis culture markers disappeared from North America around 10,000 to 11,000 BCE, the hunter/gatherer way of life continued to be strong in North America and Mexico for several millennia. Similar tools and projectile points found in the Valley of Mexico have been dated to 20,000 BCE. There were signs of volcanic activity in some of the archaeological digs, but one concludes from the data that the devastation circa 10,500 BCE may not have been as severe in Mexico as in areas closer to the ice sheet further north. Civilization in the area now known as Mexico gradually proceeded to domestication of grains and more permanent settlements as opposed to the earlier nomadic lifestyle. Signs of a divergence in culture appear first in the Vera Cruz/Tabasco Gulf coast region. The indication of this change that remains to this day is the art, such as the colossal heads and jade figurines. Due to the nature of these, many have speculated that an outside influence from Africa or Europe was introduced. Their religious symbolism continued to have a mixture of animals (some half-human/half-feline), and especially the *Bufo marines*, or toad, its forehead cleft symbolized the "cave" of creation and the concept of rebirth, their mouth as a doorway to the Otherworld. The mouth and cleft is obvious in their sculptures and figurines. The evolution of this theme is carried into the Izapan (a transition culture that perfected the calendar) and the Mayan philosophies.

Much that I read indicates a gradual developing, except for the pyramid technology and a fast-growing iconography. Was there one group from the Old World with which the ancestors had contact and suddenly there was a "quantum leap?" Maybe not. But there probably was one type of contact from which the legends of Kukulcan and Viracocha derived, and their impact ranged from architecture and astronomy to philosophy. When did this happen? Archaeologists disagree on the dating and much has been lost, but the claim by the Mayan Elders of Guatemala of a pyramidal temple 10,000 years old in the mountains there cannot be thrown out. We have already spoken of the Great Pyramid of Cholula over 8000 years old. We saw that the possible destruction of Atlantis was prior to these dates. The civilization and wisdom of the Americas are older than some may have suspected. Carlos Barrios, quoted previously, who studied with Don Pascual, a Mayan priest from the Mam Tribe, relates that the Grandfathers point to Tulum, on the Caribbean coast as one of their oldest settlements.[48]

As the ancient settlements grew, a culture with art, councils and common beliefs arose. George C. Vaillant, an anthropologist[49] and Harvard graduate, who conducted archaeological expeditions in Central America in the 1930s, gave the name 'Olmeca' to the people Friar Sahagún had called (in Spanish) "the rubber people," from the rubber trees of their heartland on the Gulf Coast. It is not known what these early people called themselves. Old Nahuatl poems speak of a land called *Tamoanchan* on the eastern sea "where there was a government for a long time."[50] (Perhaps this was Atlantis.)

The Gulf coast heartland receives about 120 inches of rain each year, which has, of course, affected the preservation of artifacts there. San Lorenzo may date to 2,000 BCE but the culture was well formed before that time. It was built upon a man-made plateau rising 150 feet above the surrounding lowlands. Eight monuments of colossal size made of basalt quarried 50 miles away were sculpted without the benefit of metal tools. These giant heads over eight feet tall, weighing up to 30 tons, have African features, and, obviously, were significant to the these people. It is speculated they may have represented ancestor worship.[51] When the city was destroyed by an unknown invader these sculptures were defaced and some were toppled into ravines and covered. Jade from as far away as Guatemala and obsidian, from the highlands, was imported for blades and bone working tools. The large quantities of *bufo alvarius* (toad) bones found here indicate that the hallucinogen found in their glands was in frequent use, probably by the shaman/rulers directing rituals. The cleft head of the toad was a motif used widely as it symbolizes a portal, the birthplace doorway, like the cleft-feature near the center of the Milky Way known to the Olmecs as the 'birthing place.' (See Illustration #2)

Milky Way Galaxy

Jaguars, of which there were many in this area, were another dominant feature in their art; some depicting half-man/half jaguar may have come from "vision" journeys with psychedelic substances. Jaguars also represented a portal to the Underworld used by shaman/priests, as did caves. Animal totems (an animal chosen to be our invisible loyal companion in this life) are a common feature in most indigenous cultures, and not only among the shamans. In healing sessions with a modern-day shaman of the Amazon/Incan tradition I was gifted from the ancestors with the animal guide (totem) of the hummingbird, after cleansing some energy from a difficult past life.

Patricia Mercier explains that this animal symbol of raw energy was considered to be an intermediary between the worlds of the living and the dead.[52] Stone stelae (pronounced, stee-leye) such as those depicting their kings (whom they considered sacred) contacting the Otherworld also give us a clue as to what they found important. Many of these columns have descriptive iconography. Jade masks made by the Olmecs have been found in the burials of the Mayan elite, implying their great value.

La Venta around 1500 BCE was undoubtedly the most powerful and sacred place in the heartland of the Olmecs.[53] The exact alignment of its fluted (to resemble a volcano) earthen pyramid, court yard and altar with the Big Dipper constellation at that time tells us that they were interested in their connection to the Pole Star or *World Tree* axis as they called it. They visualized this sacred "tree" extending from their cosmic center down to the earth beneath their king's throne. The rulers were sometimes drawn with serpent/crocodile legs symbolizing their connection to the supernatural – the Otherworld. The enormous amount of the beautiful stone called serpentine that was laboriously carried to La Venta to pave the sunken plaza in front of the pyramid (their creation place) gives an indication of the sacredness of this area. These were laid in deep layers with the top layer of square mosaics depicting a sacred entrance to the Otherworld (spirit world as we would say today). It is believed that the labor involved in these huge projects was given willingly by their subjects because their king's relationship with the divine was greatly valued.[54] The white cloth headband worn by figures carved on their stelae denoted kingship and the budding sprouts around him symbolized him as the embodiment of their world center (tree).

In this same general time period as the Olmecs, around 1630 BCE, the eruption of Thera and the destruction of Santorini along with earthquakes in Egypt rocked that part of the world. Barbara Hand Clow reports that whole islands in the western Mediterranean sank as Santorini's deep caldera sucked in billions of gallons of seawater and created a tsunami over 400 feet high.[55] Across the ocean, the Egyptians were completing their pyramids, and Stonehenge was being erected. The last violent eruption of Humphreys Peak near Flagstaff, Arizona, came around 2000 BCE. But there are many indications that the root of the Olmec culture goes back much further.

Some researchers believe that the enigmatic sculptures such as the colossal heads of the Olmecs were heirlooms of a still older culture.[56] Four of these gigantic basalt sculptures were found at La Venta and two at Tres Zapotes, a city that survived long after the former was destroyed. It is there that one of the oldest dated monuments was found; Stela C has a date that is of the Long Count calendar. From this we know that the Olmecs had an awareness of the cycles of time, a philosophy of world ages and the starting date (3114 BCE by our calendar) of the age they were in which they called the *Fourth Sun*. The basis of their astronomical observations and calculations are very ancient. The widespread veneration of the same gods, the same sculptural styles, and carved stelae, spread throughout eastern Mexico, to the Valley of Oaxaca, even to the Pacific coastal plain of Guatemala and El Salvador, indicate that the Olmec cities were not isolated. Their ideology was well established by 2000 BCE and evolved at places like Monte Alban and Izapa over the following millennia. The pyramids became more complex, some having subterranean passageways, efficient rainwater collecting systems developed, and their astronomy became more precise, as seen at Monte Alban, which was taken over by the Zapotecs around 300 BCE. Villages became sprawling towns, such a Tlatilco (dating to 1300 BCE), archaeological exploration of which exposed burial sites for 340 people; this was only a small portion of a large area that had been destroyed by brickworkers digging for clay. An astonishing variety of figurines found there reveal a philosophy of unity in opposed principles such as life and death.[57] We will examine this viewpoint of nonduality in greater detail later.

Some of the Olmec sculptures, *bas reliefs,* and paintings depict tall, thin-featured, apparently Caucasian, men with straight hair and full beards (most Amerindians have little facial hair). When we consider a culture that appeared to honor multi-racial/multi-ethnicity, we see a people of broad acceptance and a sense that there may have actually been a real person, *Quetzalcoatl,* who peacefully brought an influx of new ideas. The priesthood of the *Plumed Serpent* endured for hundreds of years, even into the Mayan time. The Olmecs and Mayans did appear spiritually and metaphysically advanced in ways hard to understand when we look at evidence from cultures further north. The hieroglyphic script and the bar-and-dot computations on stelae, as well as the observatory at Monte Alban show an advanced degree of knowledge in this hemisphere.

From Tres Zapotes we 'journey' to Chalcatzingo (600 BCE), where one monument portrays the king/shaman sitting in a cave, representing a portal to the Spirit World; the beliefs were carried onward through time and outward through the entire area. As these centers evolved, their cosmology evolved, replacing the North Pole, as the Center of their belief framework, with the "dark-rift" of the Milky Way center. Izapa was the next 'shining star' generating over eighty carved stelae [58] and miscellaneous monuments depicting scenes from their Creation Story, which had been in existence for some time in 600 BCE. The story is purely the history of their growing understanding of astronomy and the cycles of time. The legend of The Hero Twins and their father, One Hunahpu, describes, in allegory, the defeat of the false god, Seven Macaw (the Big Dipper, or Pole Star deity of the previous World Age) and reveals their knowledge that the true Center and birth place of the world was *Xibalba,* the center of the Milky Way, the 'place of awe.'

Every copy that could be found of the *Popol Vuh,* the written version of their creation story which means 'Council Book,' was destroyed by the Spaniards as they evangelized across Mexico in the 16th century. Yet it was one of their missionaries, Friar Francisco Ximénez, of Chichicastenango, who obtained an alphabetic version from one of the Lord Quichés, made a copy and added the Spanish translation. It remained in the possession of the Dominican order until Guatemala gained independence in 1830 and was placed in the library of the University of San Carlos in Guatemala City. Spanish and French versions were published in 1857 and 1861. The hieroglyphic version was among the most precious possessions of the Quiché (descendents of the Maya/Aztec/Toltec) rulers before conquest. It contained accounts of astronomical cycles and patterns of the whole *cahuleu* (the Quiché way of saying 'world')[68] or 'sky-earth.'

Izapa, the Intermediate Culture

Izapa, though far from the Olmec heartland, near the Pacific in what is now the state of Chiapas, Mexico was one of several cities that ushered in the era of the Mayans. Some archaeologists think Izapa may have been settled as early as 1500 BCE.[59] Artifacts from this time have been found all along the coast and up into the Guatemalan highlands. Many stelae of the Izapan style (always evolving in many ways) discovered near both coasts have dates extending into the first century AD, and bring us to the Mayans.

The cosmology (the philosophy and science that studies the universe as a whole, to paraphrase the New World Dictionary), beliefs, rituals and symbolism of this evolving culture is very complex; researchers/authors, such as David Freidel, Linda Schele, John Major Jenkins, Barbara and Dennis Tedlock go into great detail about these cultures, all very interesting. Much of what the Olmec/Mayans built their lives around may seem strange to us, but that was two thousand years ago!

So let's just look at the overall features. From all the references to their shaman/rulers, shaman/priests, astronomer/shamans, the belief in the powers and abilities of their 'king' or leader seems absolute. So, what is shamanism? We find examples of this kind of spiritual guidance going back to the Tungus tribe of Siberia, from whom the word originated, meaning 'the one who knows' or one with the ability to see 'with the heart' and travel to spirit worlds to bring back healing. Barbara Tedlock, one of the foremost authorities on shamanism, especially Mayan, begins her book, *The Woman in the Shaman's Body*, describing an archaeological find in the Czech Republic of the bones of a person buried under the shoulder blades of a mammoth, dating to the Upper Paleolithic age (over 40,000 years ago). The skeleton had the body of a fox in one hand, indicating that this had been a shaman, for the fox has a long history of being a shamanic spirit guide. Most interesting was the fact that this was a woman; we have learned since that during some eras shamans were mainly women.[60] Was it shaman-artists who painted the bison, horses, rhinoceri and other animals in the Font de Gaume cave in France's Dordogne Valley around this same time?

The Inuit shamans of Alaska are called 'wabinu,' or 'seeing person.' According to Mircea Eliade, "Disease is attributable to the soul's having strayed away or been stolen."[61] In the shamanic worldview, all that exists is alive, and everything and everyone is interrelated with everything else. (We will talk more about this *web* of life later.) It is the journey, or "flight of the soul" that distinguishes a shaman from other healers or mystics.[62] From all the books and workshop trainings available, we discover that shamanism is still a powerful technique today. The shaman I worked with in Raleigh, North Carolina helped enormously with the headache problems I was having and I know of one person he cured of severe allergies.

The Egyptians had a shamanic type religion in which the Pharaoh would induce an astral flight by means of hallucinogens (such as the blue water lily or opium), darkroom retreat, or meditations timed to a lunar phase in order to feel a death/rebirth (out of body experience).[63] Some say shamans are specialists in ecstasy, and that allows them to move freely beyond the ordinary world, believing in the importance of maintaining balance between the forces of the world, from the center of heaven to the center of the earth.[64] The shaman opens the 'portal' from this side; Itzamna (the ancient Mayan name for Almighty God) opens it from the other (side).[65]

We have relayed a lot about the rulers, shaman/priests, and astronomers, but we do not want to forget their ever growing numbers of farmers (growing maize (corn), beans, squash and chili peppers), potters, builders and sculptors. The artifacts that remain do not do justice to the peasants and workers. It took a huge labor force to erect the pyramids and temples in the increasing number of cities. It is estimated that in the Valley of Mexico there were 140,000 people, about 20,000 in Cuicuilco. Large quantities of ceramics and figurines have been found beneath the *pedregal* (lava) from Xictli volcano that covered that city around 100 AD.

Let's take our 'journey' back to Izapa to try to understand how their interest in astronomy correlated with their beliefs and myths. In their creation story One Hunahpu on his journey to battle the Dark Lords of Xibalba (the Underworld) chooses the Black Road (the dark rift in the center of the Milky Way) which leads there. He is decapitated (and you thought our movies were gruesome!); his skull is hung in a fork of a calabash tree (the Cosmic World Tree which stands 'by the road.') The crevice/fork the skull is placed in is a metaphor for the center of the Milky Way. Blood Moon comes along; the skull spits in her hand and she conceives the Hero Twins. (Interesting plot!) So this associates the *xibalba be* (Underworld road) with conception and the vagina of the Great Mother deity (the Milky Way). To make a very long story short: The Twins avenge *One Hunahpu's* (their father's) death and he is reborn from the cosmic birth place. (Huh?) What does *One Hunahpu* symbolize? Symbols on the stelae at Izapa are the first to tell the story: The Solar Lord (First Father, *One Hunahpu*) is the December solstice sun that is reborn as the new World Age ruler. Now, the center of the Milky Way has replaced the Pole star (the false god) as the Creation Place and the December solstice has replaced the June solstice as the beginning (creation/rebirth) point. So, 1 *Ahau* (another name for *One Hunahpu*/First Father) is the first day of the new year of their calendar. Izapa's primary ritual center is aligned to the December solstice sunrise.

The shaman/rulers at Izapa left many symbols that depict their philosophy and craft, particularly the toad motif. The toad's parotid glands (on the shoulders) contain a highly psychotropic ingredient, (also poisonous, if not properly prepared!) which assisted shamans in visionary journeys of heightened awareness. The cleft in the toad's forehead apparently reminded the 'journeyers' of the cleft in the dark rift in the center of the Milky Way where they met with the Keepers of Time. The shaman's quest for knowledge is universal and takes him/her out past our dimensions of space and time to the center and source that all religions speak of – Heaven, Paradise (Islam), nirvana, or the Spirit world to name a few. Indigenous people seem to understand the cyclic nature of cosmic time and the Mayan astronomers understood several concepts that science has discovered in recent history, such as our world (galaxy) forming or 'birthing' from the center of the Milky Way.

The heyday of Izapa (400-100 BCE) coincides with the renaissance of several of the Gulf coast cities continuing and evolving the Olmec culture. This parallels the time of Plato, Lao-Tse (4th or 5th century BCE), Buddha, Socrates, and Alexander the Great (356-323 BCE), and on to Julius Caesar (100-44 BCE), and his burning of the Great Library at Alexandria, Egypt. At this time Izapa was a large city, in the fertile coastal zone that has grown cacao (from which chocolate is made) for many centuries; the pods were used for money. It was on a major trade route both north/south and to the east. Northward is the Tacana volcano and to the northeast is Tajumulco, the highest (13,800 feet) volcanic peak in Central America. The cleft in Tacana was a sacred sighting point to which part of the ritual center was aligned to make their calendrics and cosmology obvious.[66]

As Monte Alban rose and declined, rose and fell, and Izapa became a premier site for shaman/astronomers, there were numerous other cities growing, such as Cuello with pole and thatch houses dating back to 2500 BCE. Like many other early settlements, Cuello's very ancient type pottery indicates the area was inhabited even earlier. Even in 900 BCE there was evidence of burial rituals and it had a steam bath (the oldest found to date). It continued into the first millennium, building 2 plazas with pyramids and temples.[67] Uaxactún (sometimes spelled Waxatun) was a village of wood huts in 100 BCE, but within 500 years it was a major city state in the southern lowlands, with pyramids, temples and advanced knowledge of astronomy. From a point high on the pyramid, the shaman could look directly over the top of the center stela and middle temple and see the rising sun on the spring and fall equinoxes. Across the left corner of the north temple the summer solstice sun rose, and across the right corner of the south temple rose the winter solstice sun.

The Mayan Culture

The basic beliefs, seen especially in their Creation Story and its link to the stars and sky, and ideas from their deep past, are similar throughout the Mayan region. The cross symbol represented the World Tree of the Center, the vital connection between earth and sky, man and the Creator, called *Wakah-Chan* literally meant 'raised up sky'; it was the portal that penetrated into the Otherworld, the world of spirit, or other dimensions. The shaman/king usually held this position of central conduit with his rituals of power performed for his people. First Father had "raised up the sky" with the World Tree at the beginning of creation in order to separate the sky from the earth. There were also other portals; at Palenque there were little sanctuary buildings used by the shamans. At Copan the inner sanctum of Temple 11 was adorned with the image of White-Bone-Snake (Black Transformer) where sacrificial offerings were exchanged for the *itz* (cosmic sap, 'blessed substance,' or blood, sweat, tears, semen) of the Otherworld.[68] *Itzam*, in antiquity, meant shaman - one who worked with *itz*, the cosmic sap of the World Tree. (And you thought modern religions were complicated!) Itzamna was the principle god of the Yukatek Mayans, and greatest shaman of all.

Yukatekan was the language spoken by the people in Tikal, as well as in faraway Palenque and Copan; scribes in all these places wrote in this language, and practiced the same cosmology. This unified worldview bound the many towns and kingdoms with a common vision and culture.[69] They all understood the fourfold nature of divinity; the Oneness God, *Hunabk'u*, (or Hunab K'u, in Mam dialect - *Jun Ab Ku*, which means 'unity in diversity with heart') is the Center of the fourfold cosmos – earth, air, water, fire, as well as the four directions. In traditional Mayan clans, the East is represented by *B'alam K'itzé* (fire), in the West the god is *B'alam Aq'ab* (earth), the North is *Majukutaj* (air), *I'kí B'alam* is god of the South (water).[70] Most indigenous people honor these cardinal directions with the 'creator' as center; it is a way of creating spatial order (opening a sacred space and centering the world) that focuses the spiritual forces of the supernatural within the material forms of the human world.[71] Indigenous people of North America (whom we call Native Americans in the U.S.) focused their sacred ceremonies in a similar way, utilizing the Medicine Wheel, which was a circle of stones marking a path with an alter stone in the center of the circle. From the center, the four cardinal directions were delineated, and gratitude was (and is) expressed to each of these four corners of the world upon entry to the wheel. I have only a small amount of Native American ancestry, but enjoy the meditative focus of my backyard medicine wheel. (See Illustration #3)

My personal Medicine Wheel

I ask blessings of the four 'winds', the sky, the Earth, and the center which is the Creator, as well as my power animal, the hummingbird.

As F. David Peat explains in *Blackfoot Physics*, "The term *medicine wheel* has many meanings, or rather many manifestations. (They) may have an obvious correspondence to patterns of stars and planets, but they may also be the expression of a person's dream or vision." ". . . the medicine wheel is more than a pattern of rocks, it is the relationship between the earth and cosmos; it is circular movement, a process of healing, a ceremony, and a teaching." In talking about the 'nonnormal' perceptions, powers and realities that indigenous people frequently experience or have access to, he states, "It is probably more accurate to simply say that (many) Indigenous people live their lives in a wider reality."[72]

John Major Jenkins relates how the Mesoamerican scholar David Carrasco explains that *worldmaking, worldcentering and worldrenewing* informed the beliefs of the entire region.[73] The Creation story of the Maya tells how the world was birthed from the dark rift of the 'Sky Mother,' the source, or Milky Way. Centering the village, centering the body, centering the sky was the concern of the kings and shamans, as well as each individual.[74] Centering was charted by the skywatchers; it anchored them to the center of the cosmos. World rebirth (renewing) occurred as cycles came to a close, and conditions changed; rebirth was dependent upon the proper rituals, sacrifices, and changes being made.

Over fifty of the carved monuments found at Izapa depict scenes from the creation story of the Hero Twins and the resurrection of the First Father which encode their knowledge of astronomy. Other stelae symbolically identify the crocodile/world tree as the Milky Way and the mouth of the jaguar, frog or crocodile monster as the center or dark rift of the Mother Goddess (Milky Way). Only within the past 50 to 75 years has the symbolism, hieroglyphics, and numeric system been translated so we can understand their world more fully. Harvard Mayanist Tatiana Proskouriakoff began the analyses of the glyphs; Barbara and Dennis Tedlock as well as Linda Schele and David Freidel have continued the work. Stela 11 at Izapa, we now know, portrays graphically the alignment of the stars on the end date of the Mayan calendar, and Fourth Sun (and the rebirth of the world in their belief system). Astronomer/shamans used the monuments to teach new initiates the mysteries of galactic cosmology.[75]

Uaxactun and Tikal are only 20 kilometers apart; in fact, from the summit of one of the raised temples of Uaxactun one can see Tikal. On the four corners of the entrance building opposite the big pyramid in Uaxactun are molded lords wrapped in clouds of *ch'ulel* (soul stuff) depicting holiness from the spirit world. The symbols identify this building as a *Popol Nah*, a community council house, a place where the king interacted with his people.[76] Uaxactun, however, had been established long before Tikal came under the influence of Teotihuácan, whose ideology was definitely more war-like. This is evident on Tikal's stelae and nine-step pyramid, 154 feet high, called the Temple of the Great Jaguar, built over the tomb of their great ruler, *Jasaw Chan K'awiil* (682-734 AD). We see the shifting influence as cities ascended to their peak then were either infiltrated or taken over by others. Tikal was a huge, sprawling city with a central complex of pyramids, palaces and plazas for the royalty in 378 AD when its king, Great-Jaguar-Paw decided to attack and conquer Uaxactun. The prosperity of Tikal "can be seen in the astounding proliferation of temples and public art commissioned by the *ahauob* (living gods or nobility, pronounced a-ha-wob). . ."[77] Tikal's power was challenged by Calakmul to its north in the 6th century AD, but took back its reign in 695 AD. This rivalry was in part due to their differing beliefs; Calakmul and its allies gave greater prominence to the feminine, often having joint rule of a king and queen, while Tikal's monuments reflect their emphasis on a single male ruler.[78]

"The earliest written record of a royal woman shaman in the Americas is that of *Ix Balam K'ab'al* Xook, or Lady of the Jaguar Shark Lineage . . . who lived in the city of *Yaxchilán*, in what is now Chiapas, Mexico."[79] The southern city-states had always used the blowgun as their weapon of choice, but the strength of the northern region came from their spear thrower or javelin and shield (or *pakal*). The shift in weapons and the growing tendency toward wholesale slaughter, as opposed to the previous code of taking only sacrificial captives, may be what led to the ever-widening circle of wars and collapse of the city-state kingships.

In the Old World in this general time period, the Roman Empire spread, killing many Jews in eastern Mediterranean area and Jesus of Nazareth was crucified. Emperor Constantine and the Council of Nicaea (Italy) decided in 325 AD what would go in the Christian Bible and what would be burned and banned; he established Constantinople as the new eastern capital. There would be wars, invasions, and a general decline there beginning around 376 AD. In the western hemisphere, however, Palenque and Cobá grew, reaching their zenith around 500 AD. The Mayan people, beliefs and knowledge spread throughout what is now Mexico and much of Central America. The grid plan for what would become Teotihuácan had been laid out and construction started at this time in a side valley off the central Valley of Mexico.

Cobá, 25 miles from the Caribbean Sea, had strong growth, making it a major power in the Yucatan, having around 50,000 people at its peak, and continuing to have strong alliances for several hundred years. It was the hub of an advanced road system built of stone and mortar though there were no wheeled carts or horses; the roads may have been used for religious processions and pilgrimages. Its trade routes went down the coast as far as present-day Belize and helped form its alliances with Calakmul and Tikal. A stela found there gives proof that the Mayans knew of the great antiquity of the earth; a creation date in the Long Count calendar is given in the context of a series of creation cycles.[80]

Palenque (sometimes called *Nah Chan*)[81] had roots that went back into Olmec times. Archaeologists have uncovered and deciphered much of the extensive iconography there, including the names of many of the rulers, spanning a time from 967 BCE to after 800 AD. The Pacal dynasty extended over 125 years from grandfather (*Ac-Kan* Pacal I, reigning only seven years) to daughter, who ruled three years until her son was twelve and could take the throne. This was Pacal the Great, (born 603 AD, acceded the throne in 615) who ruled for the next 68 years over this beautiful city as it attained its ascendency. His son, *Chan-Bahlum* II (reigned 684 – 702 AD), then his brother continued the plethora of architectural and sculptural works. Most famous is the tomb of Pacal the Great, leading many to speculate about connections to Egypt. Eighty-five feet below the temple, the sarcophagus strikingly resembles the mummy cases and tombs of ancient Egypt and the heavy stone lid is elaborately engraved with a scene which includes what is thought to be a side view of the king in a "transport device," for lack of a better word, to the Otherworld. He was accompanied by attendants, sacrificed at his death, as was the practice in Egypt on the death of a Pharaoh.

The legends around the beginnings of Palenque, collected and saved from burning by Friar Ordóñez in 1773, revealed that the city had been built by a people, led by Pacal Votan, who came from the Atlantic and whose symbol was the serpent. They had come from a land called Valum Chivim and, according to the Quiché Mayan book, during their journey by sea had stopped at "the Dwelling of the Thirteen." Ordóñez thought they had come from the Phoenician city of Tripoli (now in Lebanon), and others have ventured they stopped in the Canary Islands or Cuba. They came in peace and the locals allowed them marriage with their daughters, descendents of whom became the Olmecs.[82] Apparently there is some discrepancy as to which Spanish friar saved the legends. Wikipedia credits the find to de la Nada; Gilbert and Cotterell to Ramon Aguilar! Whoever saved the legends gave us another clue in the mystery of the Mayans.

The Temple of Inscriptions, the tallest structure in Palenque, with its small temple on top containing the opening to the stairway leading to the tomb, is perfectly aligned east-west and with the Temple of the Cross. Both are aligned with the setting of the winter solstice sun, assuring that when the sun arose, Pacal the Great would take up his residence near the North Star. His legacy, the Lid, assert some scholars, tells the story of the gods and goddesses of water, wind, sun, and (celestial) fire (lightning and rain), giving an abbreviated history of mankind, showing the types of destruction of previous ages.[83]

Another temple in the Palenque complex displays an elegantly carved stone skull in the white limestone at the top of 52 steps (a familiar number). Researchers/authors Morton and Thomas related this skull to the mystery of the crystal skulls, associated with the prophecies of 2012, which they pursued for many years.[84]

Palenque, of course, had its observatory room and *bas reliefs* showing a king seated on a double-headed jaguar throne (See Illustration #4, jaguar throne replica with author) receiving a crown, a throne very similar to that of Tutankhamen of Egypt. The city fell to the city/state of Toniná in 711 AD, reclaimed its primacy in 722 and continued to battle its old enemy until after 800 when all the city states came under the increasing stress of war.

jaguar throne replica with author

Though Uxmal (pronounced *oosh – mahl*, meaning 'built three times') area had been occupied since the Olmec era, it did not rise to dominance until the 8th century AD when it had an alliance with Chichén Itzá and was ruled by the Xiu family. According to legends the building of its ceremonial center, especially the Pyramid of the Magician, was accomplished easily; "all they had to do was whistle and heavy rocks moved into place," reminding us of a story told about Tiahuanaco in the Andes. Gigantic stone blocks, the legends say, had been carried through the air to the sound of a trumpet, leaving us wondering if the same could be true of the pyramids in Egypt and the possible miraculous levitation of massive stones there.[85] The House of the Governor (named for the frieze which displays the sign of the ruler) constructed on a large platform, stands at a 15⁰ angle to the rest of the ceremonial buildings. Its central doorway faces outward to the east and research has revealed that it lines up perfectly with the largest structure in what is now known as Cehtzuc, but more importantly it lines up with the southernmost standstill of Venus rising.[86] The doorway has a zodiac band that represents the sky in 910 AD. More than 350 Venus hieroglyphic symbols are positioned under rain god masks on the upper façade. An effigy carved on a nearby stela depicts a ruler on a jaguar throne surrounded by rain god masks indicating his supernatural sanction.[87] Most agree that the shaman/kings were considered vital intermediaries between the living and the spiritual worlds.[88]

When visiting Uxmal, I was most impressed with this intricate sculpturing on the buildings, especially the Serpent. Described in *The Mayan Prophecies*, the *crotalus durissus durissus* , a rattlesnake seen only in the Yucatan, has a pattern on its back like a cross within a box that the Mayans found to perfectly fit their symbolism of the sacred center and the four cardinal directions.[89] This pattern is used widely not only in Mesoamerica but among the North American native tribes as well. The Pyramid of the Sorcerer at Uxmal is unusual in that its corners are not sharp but rounded, and the steps are narrow (even for my feet!) (See Illustration #5, pyramid at Uxmal)

Pyramid of the Sorcerer at Uxmal

This pyramid, as well as other buildings here, reportedly pays homage to the rain god, Chac (or Tlaloc). The protrusions on the corners of some buildings, though they resemble elephant trunks, are postulated by Gilbert and Cotterell to be representations of the constellation Ursa Major or Ursa Minor, as there were no elephants in the New World at that time.[90] (See Illustration #6 architectural interest at Uxmal) In the highlands of Guatemala at this time the Cakchiquel and Quiché dynasties dominated the region. Uxmal came under the confederation of Chichen Itza, controlled by the powerful Itza lineage.

Ursa Major depiction at Uxmal

Let's check our timeline: we are in the mid part of the first millennium AD; so, what was happening on the other side of the ocean? The Roman Empire had crumbled; The Catholic Church had declared that any belief in reincarnation was heretical; the Bubonic plague had swept through Europe; and the Vikings were making raids on the British Isles and Russia. Meanwhile, at Chichen Itza the lower platform of the Caracol tower (the observatory) was being completed. But, we have not journeyed into the northern part of Mexico. It was not an empty place; the Chichimecs, or northern barbarians, were still mostly nomadic hunter-gatherers. Their territory changed with the weather conditions; when there was drought on the northern fringes of the city states, the farmers pulled back south to areas more favorable and the northern tribes moved further south into those previously cultivated regions. They did a little farming to supplement their diet of fruits, roots, seeds, agave (juice when water was scarce) and game, but many still lived in caves, dressed in animal skins and occasionally raided the northern outposts of society. Quietly, their numbers grew; they were to become a major factor in the fall of some cities, going on to become the Tolteca-Chichimeca and, hundreds of years later, the Aztecs as they took over and intermarried with the northern Mayans. Some accounts (after the Spanish Conquest) say the Chichimecas had no religion, while other researchers say they were devoted to the worship of the Moon. They were a major problem for the Spaniards in trying to bring the whole area under their control in the 16th century.

The Aztec legends say the gods met at Teotihuácan at the catastrophic ending of the previous age to decide who would sacrifice himself to bring light again to the world, the fifth sun. Only the sacred fire of the Creator god (Huehueteotl) remained; two gods gave themselves to the fire for the common good.[91] It is possible the city began to grow more rapidly after Cuicuilco (interpreted as 'the place of prayer') declined, then was covered by lava nine meters deep from the eruption of volcano Xitle (also spelled Xitli). It is estimated that in 40-60 AD Teotihuácan had 100,000 people, all unified in their beliefs, their expanding trade routes going out to all of Mesoamerica. The architecture, especially the Pyramids of the Sun and Moon (devoted to these deities) is beautiful, but the layout discloses an advanced knowledge of astronomy. A comprehensive mathematical survey showed that when the site was laid out, it was centered on the Temple of

Quetzalcoatl and was a scale model of the solar system, including Neptune and Pluto not discovered until 1846 and 1930 respectively.[92] Graham Hancock also points out that the alignment of these three pyramids closely copy that of three of the pyramids of Giza, thought to have astronomical significance as well as the power "to turn men into gods."

Since the massive Pyramid of the Moon is 100 feet higher than the Citadel, it is theorized that the Avenue of the Dead (about four kilometers long) may have actually served as reflecting pools (with canals that ran all the way to Lake Texcoco) which may have served as a seismic monitor (ripples would reveal tremors in the area) for earthquakes. Another anomaly found at Teotihuácan was mica; that in the Pyramid of the Sun was removed and sold by a man the government had commissioned to restore the monument. There was more beneath the stone floor of what is called the Mica Temple that is west of the Pyramid of the Sun. These massive layers are a type of mica found only in Brazil. But to what purpose? Mica is used in capacitors as a thermal and electric insulator and since it is opaque to fast neutrons, can act as a moderator in nuclear reactions. Why would they go to so much difficulty to place it beneath the floor and in a layer on the pyramid? We know that the Pyramid of the Sun acts as a permanent clock, for at noon on both equinoxes an illumination covers the shadowed west side for exactly 66.6 seconds, indicating the shaman/astronomers had a hand in its placement.[93] In the Aztec myth associated with Teotihuácan there is an old god, *Nanahuatzin*, who has to die on the pyre so he can be reborn as *Tonatiuh*, sun god of the present age. Bonfires of the material left after harvest took on a ceremonial importance as they were returning life to the sun thereby ensuring future crops.[94]

A cave underneath the Pyramid of the Sun (actually a lava tube that had been hollowed out at the end) may be an even more ancient cult center, representing the womb from which gods of the Sun and Moon, or even the ancestors, emerged.[95] There is some indication from the murals and frescoes there that, besides the Rain God, *Tlalocan* (in Nahuatl), there is a goddess called Spider Woman, who was responsible for the creation of the present universe. This correlates to Grandmother Spider of the Pueblo and Navajo creation mythology of the American southwest.[96] The gods worshipped by all these indigenous cultures – the Rain God, Sun God,

Moon Goddess, the Fire God, to name a few – were in some way all connected with maize, their staff of life.

The people of Teotihuácan produced huge quantities of razor-like blades and points of obsidian (a volcanic glass from the ancient mountain eruptions) and the rulers controlled the great deposits of green obsidian found near Pachuca.[97] The discovery of mass graves of warrior captives let us know this was not an entirely tranquil city. The area is estimated to have had a population of nearly 200,000 at its peak.[98]

What could cause the downfall of a thriving metropolis such as Teotihuácan? Some propose it was the overuse of the surrounding forests for building; others say a drought, especially severe in the Valley of Mexico, caused many of the people to leave, opening the door for the nomadic tribes to the north to come in. In the 7th century, while the Zapotecs were coming to the forefront and many of the southern cities were failing, in the Yucatan, Uxmal and Chichén Itzá flourished. Farming outposts were pushed into Chichimec territory; mining around Alta Vista was exploited, by means of slave labor, for cinnabar, hematite and crystal. Turquoise deposits were discovered in what is now the New Mexico area and brought back to Alta Vista for trade, pushing civilization further and further north. This migration eventually led to the building of Tula and had an enormous influence on the development of Chichén Itzá, though there is quite a lot of scholarly disagreement about where the "Toltec" influence originated. It appears that many of the disparate tribes of the north formed the Tolteca-Chichimeca; while sculptors and artisans from Puebla and the Gulf coast may have been brought in to construct monuments of Tula. The king of Chichén Itzá, Topiltzin (around 980 AD) was dedicated to the peaceful cult of the Plumed Serpent, Quetzalcoatl, but devotees of *Tezcatlipoca*, ('Smoking Mirror'), lord of sorcerers, took the city and forced the peaceful king to flee, legends say, to the Yucatan.[99] This increased the power of the warrior elements which eventually pushed their influence, especially that of human sacrifice, east.

The Pyramid of Kukulcan, which some say was named for the shaman/ruler Topiltzin-Kukulcan from Tula, has drawn tens of thousands to Chichén Itzá, but the profound cosmology of this merging of cultures continues to captivate the interest of many. This striking nine-step pyramid has 91 steps on each of the four sides; the platform on top brings the count to 365. On the spring and fall equinox, the shadow of a serpent slowly (three hours) goes down the northern staircase to the bottom where it connects with its serpent open-mouth sculpture, renewing the magnificence of Quetzalcoatl (Mayan for the Toltec Kukulcan), the Plumed Serpent.

Built on a rectangular pyramid, of 3500 square yards, the round Caracol (Spanish for snail, not its Nahuatl name!) observatory at Chichén Itzá must have ranked high in importance. Its narrow windows and front stairway have been thoroughly checked and the most significant alignments are the southern and the northern standstill of Venus, allowing them to predict when it would arrive as the morning star (*chak ek*, red star or great star). The shamans (who were called *tlamatiquetl* in Nahuatl) knew that the cycle of Venus was 584 days (583.92 say modern astronomers) and their giving the name of 'Four Hundred Boys' to the Pleiades constellation makes it obvious that they knew there were many more than the fourteen visible stars there. Kukulcan, when disguised as the god of wind and hurricanes, Ehecatl, was worshiped at round structures such as the Caracol.[100] They were very much aware of precession, seeing different stars as a backdrop for the rising sun on the spring equinox, due to the tilt of the Earth, as the Long Count calendar had been in use throughout a large area for centuries.

Chichén Itzá means 'mouth of the well of Itzá,' and we know that the *cenote* (Spanish) or pool there was sacred to the people. Drunvalo Melchizadek writes of speaking to a Mayan elder there and was told that people from *National Geographic* magazine had dredged the sacred pool in 1950 and found the bones of 300 Maya, who had been sacrificed or sacrificed themselves, along with thousands of crystals which the outsiders took away, much to the dismay of the Mayans.[101] In the ceremonial plaza which has a north-south alignment, is the Temple of the Jaguars and the largest ballcourt in Mesoamerica, where the myth of the Hero Twins was celebrated through sacrificial pageant.[102] The Ballcourt Stone clearly portrays the Great Cycle ending of 13.0.0.0.0 as the sun goes through the goalring (birth portal) and re-emerges from the Underworld into the new age.[103] The Ballcourt aligned with the Milky Way at midnight on the June solstices in the mid ninth century AD. A "shadow-play" on the east mural show the goal ring slowly approaching and will eclipse the "ball" on the mural around the year 2012.

The Temple of the Warriors was where the ruling council met; it did not have a king as earlier city-states. Every structure, monument and alignment encoded the Mayans' vision of the universe/cosmos. Nothing was frivolous or without meaning. Time was the great mystery and their study of the cycles within cycles gave them a galactic, "holistic worldview, based on mind as the foundation of the universe, inseparable from time and space." (José Argüelles)[104] (Reminds me of something the quantum physicists say!) They left a wealth of hidden teachings on the nature of time, cosmic cycles and evolutionary patterns. It seems that through their visionary 'journeys' and the honoring of cycles, the Mayan shaman/astronomers were attempting to anchor themselves in a galactic relationship. Argüelles also suggests "the galactic code of honor is to manifest and demonstrate harmony . . ." W. Irwin Thompson said, "As in the world view of the Hopi . . . Matter, Energy and Consciousness form a continuum."[105] The Mayans understood this; perhaps that is why both of these indigenous peoples say we are in the purification or ending phase of the Fourth Sun (or *Kin*, pronounced *'keen'* in Mayan). How could the shamans 'see' that 'portal' or 'window' in the future that would hold the perfect conditions for the rapid evolving of human consciousness? Did they perceive more from the cycles and stars than we can comprehend or recognize today?

There are numerous theories as to why the Mayan civilization, and their regional centers, along with that of the Toltec, declined. Some say the people revolted against the elitist class and increasing human sacrifice; others point to environmental mismanagement, drought or wars between surviving factions. All of which sound vaguely familiar in the 21st century, noting the situations in the Middle East, Africa and Egypt. After the fall, the militaristic Aztecs rose up in the western highlands and blended much of the beliefs, gods (their sun god was *Tonatiuh*), mathematics, writing system and hieroglyphics, much of which was chronicled in their folded deerskin or bark books. Tenochtitlan and Tlatelco were called cities of gold and stone and may have had populations of over 200,000; dykes and canals supplied fresh water from the nearby mountains. This was the legacy Moktezuma Xocoyotzin (also called Moctezuma II) received on kingship. He managed, however, to widen the gap between the nobles and the commoners by not allowing commoners to work in the royal palaces. He was killed in the battle with Hernán Cortés over Tenochtitlan, and accounts of Moktezuma'a death are varied and colorful, depending upon which of the Spanish friars or bishops was writing. The overthrow of the entire region by the Spaniards, the burning of every piece of their history that could be found, and the spread of European diseases by the conquerors have been chronicled from various perspectives in many books.

The late Mayan community of San Gervasio was probably still flourishing when Cortés landed on Cozumel. It was laid out in the classic manner with the quadrangle plaza ceremonial center, the four corners marked by the homes of its four noble families, the characteristic *Chakob* honored by the Yukatekan Maya. The Spaniard was met with a scene he did not understand: a pyramid with a three-meter-high white cross on top. *Yax che* (first tree) symbols were numerous showing that the Creation Story of the World Tree ('Raised Up Sky" Creator) was still very much honored. Cortés, of course, associated the crosses with that of the Christ in his religion. The rebirth of First Father (sun) from the crook in the ceiba tree initiated the beginning of the Fourth Sun. Those stories and the various calendars are still honored by many of the thousands of Mayans still living in Mesoamerica. The K'iche day keepers, called *ajq'ij*, still use these calendars for divination and time tracking.

While the Mayan-Toltec cultures were disintegrating and the Aztecs were becoming the first real nation-state in Mexico, Gloucester Cathedral was built in England, Chartres Cathedral was built in France and, in 1174 AD, Canterbury Cathedral was burned to the ground. The Crusades raged in Europe and Constantinople was seized; the plague spread to there, and then to Alexandria, Egypt and Europe (again!) Leonardo da Vinci was born (1450) and Columbus thought he had discovered a new world. The sun continued on its journey around the galaxy with its 'entourage,' and the scattered Mayans kept their 'days,' continued farming 'according to the stars,' and hid their calendars from the world.

TIME AND THE MAYAN CALENDAR

Cycles

Time was the Great Mystery for the Mayans as well as many other indigenous people. Many shaman-elders could 'read the Threads of Time.' The sky was their classroom; their study was rigorous. Discoveries were etched in stone and passed down orally from one generation to the next. The names may have been different with each culture around the world, but they all knew, intimately, the constellations that appeared along their path around the sun (the ecliptic). They knew that as the earth turned always toward the east, in what we know as a counter clockwise rotation (looking down on the North Pole), at dawn before the sun rose, one would see one of the twelve major constellations and about 30 days later, the next bright group of stars would precede the dawn. (Thirteen signs were noted in Asian countries, but 13 became an 'unlucky' number in the West after the edict by the Catholic Church to arrest all Knights Templar on Friday, the 13th of October, 1307. Linda Schele also notes 13 'signs' of the Maya in her research.[1]) As they noted and celebrated the longest day of the year and the shortest day of the year (the solstices), as the sun grew weaker, they prayed that it would not desert them entirely, that the 'gods' would cause it to grow strong again. As it was 'reborn' they celebrated a new year in many cultures. They noted, of course, the time when day and night were equal, a time of planting and harvesting in many places. These four days anchored their earth with the sky and their rituals and stories developed around them. It would begin to be apparent, from records kept and stories handed down, that the bright constellation in the east before the sun rose on the spring (vernal) equinox was slowly changing, moving backward from the usual monthly pattern! This must have been quite a mystery to them; we now know this is due to the fact that our planet is tilted $23.5°$, and we call it The Precession of the Equinoxes.

Whether the planet was tilted by the catastrophes around the 11th millennium BCE or earlier, perhaps when the dinosaurs disappeared 65 million years ago, we will probably never know. We do know that in the earliest indications of human history since that fateful event, people have been watching the sky. From the ancient Turin Papyrus and translations from the pieces of the Palermo Stone that have survived, which Manetho (3rd century BCE), Herodotus (5th century BCE), Josephus (60 AD), and Eusebius (340 AD) have partially preserved in their writings, it appears that the ancient Sumerians (3,000 to 4,000 BCE) and Egyptians did know about precession. Both said their calendar was a legacy "from the gods." These 'gods' must have also left a very advanced knowledge of Sirius, for the Egyptian *Sothic* calendar was based on Sirius' first rising (after a seasonal absence) just before the sun, until the next first rising to be 365.25 days, just 12 minutes longer than our solar year. This was their New Year's Day and was announced to all the temples up and down the Nile ahead of time.[117] The Egyptians had been measuring precession since the Time of Osiris (Zep Tepi, their earliest history), when Orion was at its lowest point around 10,500 BCE (a familiar year). Now, about 13,000 years (half of a full precessional) later, we will see Orion at its highest point in the sky, due to our 'tilted' angle.

Precession

How do we illustrate precession? To visualize, I stand hands on hips facing east, pretending my upper body is the earth. Now place a pole, visually, through the center of your head and down through your body to your waist; lean forward 23½ degrees toward the dawn sky on the spring equinox and you will see below your pole (which itself points to the current North Star) the very edge of the Pisces constellation and the approaching edge of the Aquarius constellation. If you swivel over a period of approximately 2125 years slowly clockwise, on that vernal equinox your pole will be at the edge of Capricorn. Although the constellations do not each encompass the exact same amount of space in the sky, over a period of 25,776 years you will come back to leaving Pisces again. I am sure the ancients did not know this fact, but we are revolving over 1000 miles per hour in our elliptical (not round) orbit 'journeying' around the sun, at 66,600 mph! In the northern hemisphere we are inclined more toward the sun in summer and more away from the sun in winter due to this tilt. Our solar system takes (approx.) 225,000,000 years to travel once around our galaxy giving us a sense of our wonderful, huge universe. From rigorous study and calculation, several scholars are convinced the Olmec/Mayan/Aztec people knew of precession. Some say the Aztec Sun Stone gives proof of their knowledge because the five ages have approximately 5200 years, equaling 26,000, which is about the time elapsed in one full precession.

The K'iché (Quiché) Mayan creation myth, illustrated on Stela C at Quiriga says, "The Maker/Creator allowed the sky to be lifted from the Primordial Sea. First Father (One Maize Revealed) raised up the *Wakah-Chan*, the World Tree (the Milky Way) and gave circular motion to the constellations," and the story goes on to personify every major star group in the night sky. Sky and Earth were quartered into four parts. But I can find nothing that proves the Mayan Four Suns (Ages) were of equal lengths of thirteen baktuns each. In fact, from the descriptions, it would seem that the First Sun occurred millions of years ago when our ancestor, the upright hominid known as *Australopithecus afarensis* did not make the 'cut' and was replaced by *A. aethiopicus and A. africanus*: "On the first try (they got) beings that had no arms to work with and could only squawk, chatter and howl. . ."[3] "The divine fathers created the (next) generation . . . out of mud, but they fell apart and returned to the earth"[4] (end of the Second Sun). We can postulate this may fit into our geological timeline when North America had the massive eruption of the Yellowstone caldera, which may have set evolution back a bit! The *Popol Vuh* relates that the gods consulted Xpiyakok (First Mother) and Xmukane (her husband, the Maize God), whom they called the Grandparents, to oversee the new Creation, the Third Sun, and this time the people were made of wood, but they were idiotic, brainless and did not remember their Creators, so they were destroyed by a monstrous rainstorm and flood. Does this sound like *Homo habilis* or *Homo erectus*, who disappeared over a million years ago? Or another version: the humans became evil and filled with pride, so they were annihilated when black storm clouds formed and the earth was inundated by a great flood. Homo sapiens neanderthalensis?

Another translation indicates the Third Sun ended because the sun's rays were blocked by smoke and flames. Almost sounds like another outbreak of volcanoes such as Yellowstone's last eruption 630,000 years ago. This is where the theory becomes very speculative. Some say it was in this period that the earth was moved to its tilted position. 67,000 to 74,000 years ago the Lake Toba, Sumatra, Indonesia, volcanic eruption reportedly killed 90% of the world's population and plunged the world into a 1000-year cold snap.[5] This volcano (caldera 100 km. long, 30 km. wide and 1,666 feet deep) has been called Yellowstone's "bigger" sister and was three times the size of Yellowstone's latest eruption. It lasted two weeks and covered parts of Malaysia 30 feet deep in ash, central India twenty feet deep. This same caldera had two previous eruptions 700,000 and 840,000 years ago. So the prehistory legends passed down through the generations have much to work with in the way of Earth catastrophes.

We have seen evidence of the extent of the damage to land and life in the 11th millennium BCE – 13,000 years ago. Studies as diverse as tree rings, plant pollen and oxygen isotopes reveal that a little over 5000 years ago there was, indeed, a severe climate disruption leading to drought and famine in the western hemisphere. The time around 3100 BCE also saw the beginnings of Stonehenge, the great stone circles in Cumbria, and Stennes Stones of Newgrange, in Ireland. The Dynasty of Egypt began and the Maya dated the beginning of their Fourth Sun at August 11, 3114.[6] Geoclimatologists found a period of climate change around this time, revealed in a 'sulphate deposit' in Greenland ice cores that suggests either cometary impact or volcanic eruptions (or both). There is evidence of flooding in Mesopotamia, Brittany and in the area now the Navajo reservation in North America.[7] We do know that this Mayan Fourth Sun, ending on 12/21/12 is 5,128.76 years long or approximately one-fifth of a precessional cycle.

If we go back from August 11, 3114 BCE about 5000 years we have 8114 BCE, which would be the end of the Second Sun, if counting an equal number of years for each age. We have no conclusive evidence of a catastrophe at that time. So, it is hard to extract a comparison between the Aztec calendar and the Mayan Long Count, though they came from the same traditions. Some say that the Aztecs had lost some of the technical erudition that the Mayans kept in their shaman to student initiations. The Creation Story going back to the Olmecs appears to be more about the creation of the universe and humans and the Long Count calendar about cycles and ages.

Several researchers have tried to correlate the various calendars to each other, to our calendar and to what we think we know about the history of our universe and it is interesting reading but few have found analogies that really make sense. I like Geoff Stray's quote in *Beyond 2012* about these: "It seems that people are so wary of throwing out the baby with the bathwater that they have poured all the water into one bath, inadvertently drowning the baby."[8] In others words, we may be losing sight of what the Mayans wanted to communicate to their progeny and the world.

In the Mayan creation story, Xmucane grinds the yellow and white maize (corn) together and models the first four beings. But the gods were alarmed that they had perfect knowledge and divine powers, so they decided to put a fog over their eyes.⁹ The end of this Fourth Sun will be December 21, 2012 at 11:11 UTC, so most scholars report. 13.0.0.0.0 is 4 Ahau in the Tzolkin (the 260-day calendar) count that is maintained by many of the modern Maya daykeepers to this day. Carlos Barrios, who studied with Mayan spiritual elders, calls the Fourth Sun *Kajib Ajaw* and says it has been governed by masculine energy; "we believe we are all-powerful, we worship reason and materialism, and we have become slaves to our own marvelous innovations. We must stop and remember what is truly valuable or we are heading toward the destruction of Mother Earth and all of humanity."¹⁰ He believes the Fifth Sun (Job Ajaw) will be governed by a balance of masculine and feminine energies and humans will ascend to a more harmonious spiritual level.

How does precession enter into the study of their calendar? Moreover, why did they care what would be happening to their descendants in 2012? At times it appears that many people now do not consider what will happen 2000 years in the future the way the Mayans did. I believe their calendar was based specifically on this end date because it was not only the end of the Thirteen Baktun cycle in which they were living, but the end of an even greater cycle, and perhaps an even larger Great Cycle! Their visionary insight was much more than 'technical training' in astronomy. The knowledge gained on the 'journeys' the shamans took on 'The Threads of Time' was relayed to their leaders; the kings commanded that the date and the message be cut into stone by the stonemasons. It was important to them.

John Major Jenkins' research indicates that the Mayans were aware of precession (our planet's slow 'wobble' due to the tilted axis that causes the vernal equinox to show a different constellation approximately every 2125 years just before sunrise). He quotes from the dissertation thesis of Dr. Michael Grofe and his work with the Maya Dresden Codex that indicates the Mayans had a precise calculation of the rate of precessional drift. Dr. Grofe's deciphering of Monument 6 at Tortuguero and of the lunar and solar alignments with the 'dark rift' (center of our galaxy) at Palenque also helped him determine this. Further verification of their knowledge of precession was the calendrical bones from burial 116 at Tikal, first brought to light by B. McCleod and M. Looper. This research led them all to believe that the Mayans were very aware that the rate of precession was one degree every 72 years.[11] The alignment of many of their buildings with the "womb" (center) of the universe indicates they were aware this 'dark rift' was coming lower and more in line with the December solstice sun, and makes us appreciate their architectural wisdom and knowledge of astronomy. Remember, the dark rift meant birth and transformation and any date in which one of the major planets aligned with this 'center' was meaningful to the Mayans. As the *Popol Vuh* (sometimes called *The Book of the Dawn of Life*) says, humanity is transformed at each successive Age to more fully honor . . . the creator. Stories such as this carry a perennial wisdom and speak in a numinous space/time that was, is, and will be.[12] (John Major Jenkins)

The Calendars

*N*ajt is the Mayan word for space/time/frequency; everything that we know as physical reality (dimensions 1 through 4), according to Carlos Barrios, arises from a giant macrospiral (near the Pleiades).[13] He further explains that the confluence of energies (vibrations, influences) from the four corners of the universe are what support the elements (Fire, Earth, Water and Air) and provide information on . . . our moment of birth as well as the terrestrial currents, our birthplace and the traditions of our people. We will look at astrology a little in the next chapter; suffice it to say, that the Mayans took the influence of energies coming from the planets, stars and constellations on one's life very seriously. Barrios' *Book of Destiny* includes charts where one may find their Mayan birth sign, one of the twenty named days. My sign is designated by the hieroglyph *B'atz'*, which means Thread of Destiny or infinite time and unity. "This sign symbolizes cosmic phenomena and original wisdom. *B'atz'* is the deity that created Earth and Sky, the Creator of life and wisdom. . . (it) presides over the future." Barrios also relates that some of the twenty Mayan calendars have not been revealed because the time is not right; his information comes from the modern Mayan elders and early texts. The ancient *Ch'umilal Wuj* or Book of Destiny, based on the *Cholq'ij* calendar (tzolkin in some dialects), contains "the ancestral technique" for interpreting the "different meanings of the twenty days, or *nawales*, the year bearer, and the Mayan Cross or Tree of Life."[14]

Schele and Freidel explain, "We count with our fingers and base our numbers on units of ten. The Maya counted with the full person, both fingers and toes, and based their system on units of twenty." "We mark the passage of decades, centuries and millennia; they marked the passage of 20-year cycles, which they called *katuns*, and 400-year cycles (20 x20 years), called *baktuns*." Their book, *A Forest of Kings*, has a great illustration on how the *Tzolkin* and the *Haab* or vague year, meshed together like cogs to make the Calendar Round, the 52-year cycle.[15]

Though the *tzolkin* (in some dialects *Cholq'ij*) calendar of 260 days (based on the human gestation period and/or the phases of Venus which is 258 to 263 days depending on how it is counted) had been in use a long time as well as their solar calendar (*haab*) of 365 days, it was where these two calendar wheels meshed that was important to the Olmecs, Izapans and Mayans. This intersecting (we now call it the Calendar Round) happened every 52 solar years and the New Fire ceremony at that time was paramount. The Pleiades constellation had to pass through the zenith (precisely overhead) at midnight to ensure 'continuation' and the beginning of a new year. It appears the Long Count calendar developed alongside these, but since the recording of dates does not show up in the archaeological record until the first or second century BCE, we have only the much older pictorials to go by.

Some scholars believe the Long Count goes back to the 6th century BCE and was perfected in the era of Izapa (400-100 BCE). All Mayan calendars are based on multiples of 20: a *tzolkin* was 260 days (the length of human gestation, counting after the first missed period), or thirteen months each having twenty named days and the *haab* consisted of 18 months of 20 days (plus *Uayeb*, the five unnamed, unlucky days) was called a *tun* (a solar year). 20 *tuns* was a *katun* (approximately 20 years); multiply by 20 again and you have a *baktun* (144,000 days) or almost 400 years. Thirteen *baktuns* is almost 5,200 solar years, which ends One Creation (or Sun) and begins another. One Sun or Age was approximately one-fifth of a precessional cycle, which is 26,000 years.

The Mayans started their present Long Count from the destruction at the end of the Third Sun (another cataclysm); and the beginning of the Fourth Age (13.0.0.0.0 in Mayan notation) was on or about August 11, 3114 BCE in our Gregorian calendar correlation. December 21, 2012 is the approximate end date of this 13-baktun cycle, the Mayan Fourth Sun (end of the Aztec's Fifth Sun and the Hopi's Fourth World). It is difficult to tie the Long Count calendar to the *Popul Vuh* creation story for that legend appears to recount the history of the evolution and development of humans from their earliest beginnings going back tens of thousands of years. The Mayans saw every cycle as part of the great macrospiral (*Kan*).[16] Cycles within cycles. Their predictions were based on events that happened at the turning of previous cycles, such as at the end of their Third Sun, which, of course, they related to what was known or seen in the sky at that time. *As above, so below.*

To recap: The Long Count calendar is based primarily on multiples of twenty, just like the *haab* and *tzolkin*:

NUMBER OF DAYS		TERM
1		*kin* (day)
x	20 named days	*uinal* (month)
x 18=	360	*tun* (year)
x 20=	7200 days	*katun* (19+ years)
x 20=	144,000 days	*baktun* (394.5 years)*
x 13=	1,872,000 days	13 *baktuns*= 1 Cycle
=	(5,125 years)	1/5 of a precessional
1 precessional =		approx. 26,000 years or 1 Great Cycle

*From *Maya Cosmogenesis* by John Major Jenkins[17]

The Great Cycles show our collective unfolding as a species, as we can see from the legends of the first three Suns. The Maya considered the end of a 13-baktun cycle to herald a major World Age shift. They knew that when the December solstice sun crossed the center of the Sacred World Tree in the cosmic 'womb' (the center of the galaxy) humanity would reach its next phase of spiritual development. The Mayans understood that cultures, like humans, grow and mature.[18] In the Mystery Schools of the Greeks and Egyptians precession and world ages were part of the secret teachings and those initiations were taking place in those countries around the same time the Olmec/Maya were perfecting their creation story and calendars.[19] The cross-referencing of *haab, tzolkin,* and Long Count calendars on documents, stelae, monuments and buildings allows us to be sure of the translation of their dates to our Gregorian calendar. We see the importance of the numbers 20, 13, and 52.

The Stars

The Mayan's complex cosmology included the moon, Venus, Orion and the Pleiades, but the rebirth of *One Hunahpu (or Hunab Ku)* on the December solstice was the senior sign because this deity represented the sun. In the *Popol Vuh*, as we have noted, *One Hunahpu's* resurrection signals the beginning of a new World Age and is still celebrated by the Chamula Maya.[20]

This 2012 rare event is not just the end of a world age, but the ending/beginning of a Great Cycle, one that only happens every 25,625 years. The Pyramid of Kukulcan at Chichén Itzá, besides the 'shadow play' of the serpent descending the steps on the spring equinox, has another alignment pointing to 2012. Sixty days after the equinox, the zenith passage of the sun there will combine with a solar eclipse on May 20, 2012 (tzolkin day sign *Chicchan* which means serpent) and further emphasizes the serpent's rattle (*tzab*) which is the same word used by the Yucatec Mayans to designate the Pleiades. To further elaborate the symbolism, the little face on the back of the rattlesnake (serpent) is identical to the glyph for *Ahau* or sun.[21]

In the *Popol Vuh*, Zipacna escapes and the 400 youths that tried to kill him are removed to the sky and become the Pleiades. Sometimes referred to as 'The Seven Sisters,' we know and the Mayans knew the Pleiades has many more than seven stars and have been venerated not only by the Mayans, but indigenous people in Siberia and Australia, for thousands of years, some even claim Pleiadian origins. The Mayan symbolism interrelates to form a complex cosmology.

One researcher who has written extensively about the Pleiades is Barbara Hand Clow, who claims the Pleiadians contacted her while she was in an altered or trance state and transmitted information concerning the nine dimensions of consciousness. These dimensions correspond to nine physical parallels in (1) the iron core crystal in the center of the Earth, (2) the magma or liquid elements that surround the core, (3) Earth's crust, time/space reality, (4) collective consciousness, (5) love and creativity held in the Pleiades, (6) morphic (specified form or shape) fields and geometry accessed through star system Sirius, (7) galactic information/light held in the Andromeda galaxy, (8) cosmic order from Orion, (9) the Milky Way galactic center (time). She believes the Mayans, especially the shamans, understood this multidimensionality.[22] I have explained this in my terms and hope I have relayed it properly; her work is complex, but interesting. She believes the Pleiadians are overseeing the evolutionary awakening on Earth. (I warned you we were going to stretch our minds to all kinds of available information!)

Barbara Marciniak is another writer who channels information from the Pleiadians and her book, *Earth,* gives some valuable clues to the evolutionary aspect of Earth and humanity. She reveals that as the Mayans called themselves Day Keepers, the Pleiadians refer to them as the Keepers of Time and that "the Maya laid the groundwork . . . for events that are to transpire (in this time)." This source tells us that the Egyptians, Incas, Native Americans, as well as the Atlanteans, left in place other types of information for this time. Like Atlantis, "versions of your world will find solutions and move into the Golden Age, just as *versions** of your world will be destroyed."[23] Ms. Marciniak says the purpose of the Maya was to establish a paradigm for the future, to hold open portals of energy through collective consciousness that support other realities.[24] So, we notice an ongoing connection to this faraway group of stars. It almost feels as though these older cultures believed (or hoped) that humankind would in this new era be able to 'see' beyond the four dimensions we now experience as our whole.

Every 52 years when the Pleiades reached its highest point (zenith) at midnight, the Mayans, Toltecs and Aztecs would extinguish all fires and conduct the New Fire ceremony, relighting the fires and commencing a new beginning. This also was the year that their 260-day calendar and the 365-day calendar meshed and came back to their starting points. Morton and Thomas (*The Mystery of the Crystal Skulls*) point out from computer generated calculations that just before sunset on December 21, 2012,

Venus will sink below the western horizon and the Pleiades will rise on the eastern horizon, symbolizing the death of Venus and the birth of the Pleiades.[25] What else will happen that year?! Let us 'journey' further.

We have seen that Venus was one of the principal alignments at Chichén Itzá. The Venus Table of the Dresden Codex (one of the few Mayan documents preserved from the Spanish mass burnings of the 16th century) gives us a glimpse of their worship of this planet as well as the extent of their knowledge of mathematics. They knew how the Venus cycle (584 days) related to the lunar cycle of 29.53 days (another x 20). Venus was recurrent in their sculptures, and was often referenced in dates that correspond to its heliacal (with the sun) rising. Their science and religion were closely related through the stars; the cycles of heaven and the gods represented a kind of perfection for which mere mortals could strive. What happened in the sky mirrored what happened on earth. This example of order was a stable pillar to which they could anchor their minds and souls.[26] (Anthony Aveni) "As above, so below. As within, so without." The Mayans believed that everyday life and the natural world were meant to function in harmony as a Whole.[27] They were convinced that humans were inseparable from the natural and cosmic cycles.

Terence and Dennis McKenna, after study with indigenous shamans in the Amazon during which they took numerous hallucinogenic 'journeys,' wrote that "time travels in cycles or waves across the universe" and felt they had found a way to plot the complexity of change within time, which they called 'timewave zero.' Terence McKenna worked with mathematicians and nuclear physicists to better explain his theory, showing that all the 'waves and subwaves' would peak together, signifying a dimensional transition.[28] Gregg Braden worked with McKenna's ideas of events becoming the seeds of future conditions and having been a Senior Computer Systems designer for Martin Marietta Aerospace he was well qualified to study it. As he says, "the scientist/shamans of our past created a beautiful bridge between the worlds of sensual beauty and time's cycles . . ." "The secret (is) all about recognizing the way cycles . . . play out in our lives."[29] These 'great waves of energy that pulse across the universe'[30] that we call Time are the cycles that the Mayans could perceive in their 'Otherworldly' journeys. This cycle ending and the transition into a new age is a process. Though it seems our planet and solar system are moving at a great speed, this galaxy is so large our alignment with the center of it began several years ago and will only be at a mid-point in 2012. Does this mean we are at the 'end of time'? Do not underestimate the power and significance of human consciousness as well as that of Infinite Intelligence!

Perhaps, we are coming into a space/time where/when we will 'see time differently.' You can now observe that our 'journey' has taken us around to the effect these cycles, 'timewaves,' or planetary alignments may have on us as the 'captive' travelers on this planet, and to the question: "Can we influence the effects of these cycles, especially one so grand as to be called the end of a Great Cycle?" Hint: reread the last sentence of the previous paragraph. The Mayans called the time period beginning on December 22, 2012, the Fifth Sun and the beginning of a new world age and a new Great Cycle. It is obvious they believe we shall all continue on into a New Age. What do our astronomers see in the sky?

ASTRONOMY AND ASTROLOGY

Our Sun

We have tried to see the sky through the eyes of the Mayans of the first millennium AD and their ancestors. We found their astronomer/shamans' observations very perceptive considering the technology of that time. Let us 'journey' to our own astronomer/scientists to see what they have to offer about our 'sky.' Shall we start with our sun? We would not be here without it! It has been considered a 'deity' worldwide since humankind evolved. They may have thought ours was the only one, but we now know that there are hundreds of billions in our galaxy alone! The SOHO spacecraft has been studying the sun since 1995 and though it cannot see through its opaque surface, it has been sending sound vibrations back to helioseismologists (how's that for a BIG word!) These signals have helped them recognize the processes going on inside its four layers. The sun's matter is not solid, liquid or gas but plasma, called the fourth state of matter. Plasma is not stable like the matter we are familiar with, which has atoms of balanced negative and positive particles.[1]

So, what is plasma? Wikipedia describes plasma as an electrically neutral medium of free positive and negative particles with an overall charge of roughly zero. Although these sub atomic particles are unbound, they are affected by each other's fields.[2] Plasma contains positive ions and negative electrons and responds strongly to electromagnetic fields. Those of you who are familiar with all these terms bear with me while the rest of us try to understand all this! Plasma does not have a shape or volume unless enclosed by a magnetic field in which case it may emit light in the form of filaments such as is seen in plasma TVs, fluorescent lights, neon signs, plasma torches, and many other research areas such as fusion energy research. Signs of plasma are also seen in lightning, ball lightning, polar auroras, and in our ionosphere just to name a few. Within plasma, positive and negative particles move around freely like a fluid, even though they may be compressed to well beyond the density of most solid matter we find on Earth. Over a million megatons of pure light energy (photons from the full electromagnetic spectrum) are released every second from the sun.[3]

Our sun is not a solid body; there is no edge or hard surface to it. It is 864,000 miles in diameter; more than one million earths could fit into it. Nuclear fusion in the core reaches around 28 million degrees Fahrenheit, while the surface is a balmy 10,000° F. Under this massive heat and pressure, two hydrogen atoms are slammed together at such high speeds they form a helium atom releasing high-energy photons or gamma rays, which accounts for 98% of the sun's energy.[4] It is estimated the Sun will burn through its storehouse of hydrogen in about 5 billion years. The radiative zone of plasma around the core has the density of gold and the photons leaving the core may take a million years to get through this layer but mostly come out in a safe spectrum of light.[5] The magnetic field of the sun is not like Earth's, with a simple north-south alignment. It also has an equatorial quadripole with alternating polarities, and because the equator is turning faster than the poles, its magnetic flux lines get wound up into loops.[6] These, twisting throughout, pull streamers of gas far into space and occasionally blast huge plasma storms outward.[7] The sun's gravity, of course, is what holds our planetary system together, and extends all the way out to the Oort cloud on the edge, 200,000 AU away. AU is the abbreviation for astronomical unit and one unit is about 92,955,807 miles, or about the distance from the Sun to the Earth.

Sunlight is visible electromagnetic radiation and the amount reaching us in one year contains over 20 times the energy of our entire reserve of coal, oil and natural gas.[8] Sunspots and solar storms cause small variations to this constant. What is electromagnetic radiation? Well, we know about visible light, infrared and ultraviolet are just below and above that, below infrared are radio waves, and above ultraviolet are x-rays and gamma rays. (See Illustration #7, Electromagnetic Spectrum) I mention this here because we will be examining plasma and electromagnetic radiation later as it applies in our 'journey' to understanding some of the Earth/human changes around 2012. Einstein applied Max Planck's quantum theory to show that all forms of electromagnetic radiation, including light, travel in tiny bundles of energy called photons and can behave either as a wave or a particle. Virtually every point of space is saturated with electromagnetic radiation, including the light from millions of stars.[9]

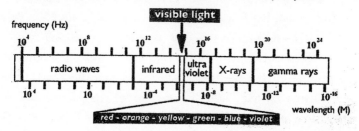

Electromagnetic spectrum

Other phenomena that we need to understand are solar cycles, sunspots and 'sunquakes.' There is a periodicity in the appearance of sunspots that is approximately eleven years from peak to peak. Scientists noticed that there are bright solar flares near sunspots, followed by geomagnetic storms on the sun concluding, after further study, that sunspots have magnetic fields 100 times stronger than normal.[10] The sun's magnetic fields have become a major area of study, because they affect the earth strongly. These fields appear to be generated between the radiative zone and connective zone, due to their differing rotation, as well as the solar equator rotating at a slightly faster pace than the solar poles. So these magnetic field lines become twisted and tangled and leave the sun's surface in giant loops. At the base of each loop is a sunspot, sometimes many thousands of kilometers wide. The magnetism keeps a 'lid' on hotter gases at the center of the sunspot, so the energy flows out around the edge. Solar plasma follows the arch of the magnetic loops and when numerous and come in our direction, these coronal mass ejections can cause magnetic storms in our atmosphere, like the huge one in 1958 and the one in 1989 that knocked out electrical power to five million people in Quebec. As these arching magnetic loops from two sunspots cause the plasma to get hotter, up to 212 million degrees F, they break, reconnect and release huge amounts of energy causing a solar flare.

There were several unusually powerful flares during the last solar cycle: March/April, 2000 was supposed to be the peak of the last cycle, but on April 2, 2001 a solar flare broke all previous records, going above an X-20, the highest ever recorded at that time, causing the aurora borealis to be seen as far south as Mexico. In November of 2003, in the space of twelve days we had a record-breaking nine major solar eruptions and the largest one ever recorded, estimated to be over an X-35; scientists in New Zealand concluded that it was an X-45.[11] An X-2 flare on New Years Day and an X-7 later that month in 2005 were small in comparison, but set the scene for the record setting hurricanes of that year. The flares of 2005 were exceptional in that the proton storm reached the Earth fifty times faster than normal, about 75,000 kilometers per second, with a direct hit. (By comparison, light travels at 300,000 kilometres, 186,000 miles per second.) An X-17 flare in September caused a blackout of many shortwave, CB and ham radio transmissions in the Western Hemisphere and flares continued with nine more that month.[12]

In March of 2006 the flares suddenly stopped, NASA declared the end of Cycle 23, and NCAR (National Center for Atmospheric Research in Boulder, CO) predicted that the return of these sunspots in the next cycle would be amplified, thirty to fifty percent higher.[13] When 2008 had sunspots, it was surmised that there was an overlap in the two cycles. The projected next peak was pushed forward to 2012, which will be solar cycle # 24. The radiation from strong flares could be deadly to astronauts if they were to take a direct hit, especially a proton storm travelling as fast as those in 2005. These charged particles can confuse compasses and Global Positioning Systems, which is especially dangerous to those used by commercial airlines. But what about us on Earth? Are we seeing the effects of extreme solar cycles by the increase in crime, cancer and other diseases? We have always joked in my family, when having a particularly trying day, "There must be spots on the sun!"

The sunspot cycle, from peak to peak, has been placed at 11.87 years and, interesting correlation, the orbit of Jupiter around the sun is 11.862 years. Coincidence or contributing factor? Our satellites monitoring 'solar wind' are our 'early warning system' for strong proton storms coming toward Earth and these might give as little as 30 minutes advance notice! Solar wind is just another way of saying, 'the stream of charged particles (negative electrons and positive ions) constantly flowing outward from the sun.' The total number of particles carried away per second is a phenomenal number, yet only .01% of the Sun's total mass has been carried away by solar wind. It is thought that the extreme acceleration, especially at the poles, is related to the 'coronal holes' or open magnetic field lines in its atmosphere. Coronal mass ejections (CMEs), fast-moving bursts of plasma, associated with solar flares, impact the Earth's magnetosphere and temporarily 'deform' the Earth's magnetic field. This is called a geomagnetic storm and there are many types and speeds that originate in our sun. Our magnetosphere is a 'bubble,' for lack of a better phrase, with a trailing extension or 'tail,' around our planet, with the leading edge, called the 'bow shock,' taking the brunt of the solar wind storm.[14] If we did not have our strong magnetic field, our atmosphere, like Mercury's and, to some extent, Mars', would be ripped off by the solar winds. Our moon has no magnetic field so its surface is constantly bombarded by the solar wind. This constant outflow from our sun, however, forms a somewhat protective 'heliopause' outside the orbit of Pluto to defend our solar system from interstellar and galactic assault. We are entering a period where we may know just how protective this 'shield' is. There is an excellent diagram of the heliopause at wikipedia.org.

The Milky Way

Imagine our galaxy as a spiraling disc that is actually quite 'thin' (7,000 light years thick) compared to its (more or less) 100,000 light year diameter. (Think frisbee, slightly thinner.) If you visually slice that in half horizontally, like a bagel, you will locate its 'equator,' or horizontal midline. Our solar system is on one of the outer 'arms,' called the Orion Cygnus arm, and over a very long period of time (some scientists say 64 million years, others say 37 million years), 'journeys' above and below this midline plane. It takes us (our solar system) roughly 225 million years to go around our galaxy once, 'cruising' at about 155 miles per second. The Milky Way (with its 200 billion stars) is also 'travelling' (with all of 'us' aboard) and has its own shock wave and galactic wind of superheated gas and cosmic rays. As our entire solar system is now coming above the mid plane of the galactic equator, we are much more exposed to that assault, as well as the possible increases of our own solar wind. This places us exposed to more asteroids, meteors, cosmic 'dust' and radiation, by a factor of 10 some scientists estimate. Dust might not seem like a threat, but it appears that the sun's magnetic field weakens during its periods of increased activity, thus allowing two to three times more interstellar dust into our solar system. We find a number of factors falling into the 'unfriendly' category, as far as the Earth is concerned, and we have not even looked at the magnetized plasma theory, solar megacycles or spacequakes!

Dr. Alexey Dmitriev, professor of geology and chief scientific member of the United Institute of Geology, Geophysics and Mineralogy in Nvosibirsk, Siberia, has caused many 'ripples' in the 2012 investigations. He has pointed out the anomalies and weakening in the Earth's magnetic fields. Dr. Gauthier Hulot of the Institut de Physique du Globe de Paris and the Danish Space Research Institute compared measurements taken by the Danish Oersted satellites with earlier ones and found the same phenomenon, plus an area below South Africa that is reversed from the normal magnetic alignment.[15] In 1996 it was discovered that Earth's crystal core revolves faster than the molten outer core generating our magnetic field. It is speculated that within the currents of molten iron there are huge whorls and vortices, that these are causing the anomalies, and may increase and further weaken the overall field. From a study in 1986 reported in *Nature* magazine, scientists found that magnetic reversals on our planet have been linked to sharp drops in Earth's temperature.[16] But we do not appear to be dropping! Yet! Dmitriev also reports that Jupiter's magnetic field has doubled and now has "auroral anomalies."[17] (Jupiter has always run blocking/interference for Earth with incoming asteroids; we know it took a hit for us in 1994 with the comet Shoemaker-Levy-9.) Dmitriev attributes all these to our transit through a magnetized band of plasma which is building up on the heliosphere and then breaking through into the interplanetary domains.[18] Some call this an interstellar energy cloud and say the radiation from it is increasing in our ionosphere, 60 to 100 miles above the Earth, causing changes which Dmitriev says are leading to an evolutionary change or transformation, just as the Mayans and other indigenous people have predicted. The Mayan Grandfathers call this "a return to the Beginning."[19]

Many of these events seem to be connected to the precessional cycle, or half cycle of 13,000 years, in which the North Pole points either toward or away from the galactic center. Dr. Paul LaViolette has added much to the understanding of why the ancients may have tried to warn us of a possible upcoming event. He says that the galactic core is not a black hole, but a massive object that explodes periodically, specifically every 26,000 years, and possibly, every 13,000 years. It is interesting that the time period matches that of our precessional cycle. LaViolette proposes that this explosion sends out galactic superwaves of cosmic radiation. It appears our astronomers cannot see this happening because of dust and the intense light from the millions of stars within the 30-light-year center. Alexei V. Fillippenko (Proc. Nat. Acad. Sci. 1999 96:9993) appears to agree with LaViolette in noting that observations made with infrared wavelengths point to the center of our galaxy being dominated by a single object with a mass equal to millions of our sun.[20] Other scientists maintain that our galaxy has a Black Hole in the center like many other galaxies. They have named this area Sagittarius A* (pronounced A star). Whatever is there, we know it has the capacity to destroy stars and also to "fling" them out of the galaxy at speeds of 1.6 million miles per hour, as scientists discovered in 2005 and fifteen more since then.[21] Ice core examinations have shown that almost 13,000 years ago (the time of our aforementioned catastrophe) there were spikes in acidity, indicating cosmic radiation. 12,850 (There is that familiar date again!) years ago our polar axis was facing the galactic center and the planet underwent a solar flare cycle that was hundreds of times more intense than recent years.[22]

Other Dangers

In 1998 there was a bombardment of x-rays and gamma rays (from a collapsed star) upon the Earth, and B. H. Clow theorizes it may have started a change in the evolution of human consciousness.[23] Scientists have questioned whether this type of event may have been responsible for the some of the major extinctions on this planet in the past. I suspect that these blasts may have come with or precipitated other consequence such as 'loose asteroids.' Clow also reports in *The Mayan Code* that scientists say the expansion of the universe is speeding up; all I can find is that there is still quite a bit of controversy on that subject. A balance between Matter and Dark Matter seems to be at the heart of that discussion, and work on the 'superstring theory' may give some insight into dimensions beyond our known four, as well as that mysterious fifth essence called *quintessence*, which has something to do with vacuum energy, anti-gravity and this dark matter - all way above the level of our present inquiry!

NASA reported on their website that on July 27, 2010 researchers using their fleet of five THEMIS spacecraft have discovered a form of space 'weather' that they call a "spacequake." This is a tremblor in Earth's magnetic field that begins in the "tail" of the magnetosphere (remember the magnetic 'bubble' around us?) that is stretched out behind us by the million mph solar wind. It can become so stretched that it snaps back like a rubber band and plasma hurtles toward Earth. These jets of plasma crash into the geomagnetic field about 30,000 km. above the equator with plasma vortices (huge whorls of magnetized gas) on either side generating electrical currents, sparking auroras and disturbing communications.[24]

Russian biophysicists studying the human 'energy body' with an electron microscope found that within the body's magnetic field is a 'biological plasma body.' This 'body' is charged up particularly by ionized air and is effected by the 11-year solar cycle, the moon, thunderstorms, strong winds, illness, emotions, thoughts of others, and changes in the planets![25] (Does that refer to astrology?!)

One other factor in our 'safety' while 'journeying' around the universe is the fact that Earth's orbit around the sun is 'eccentric.' (And I thought that only applied to certain humans!) The reason this is important is that even a small change in solar output or reception can have a significant effect on many living organisms here. The somewhat circular path we take around the sun varies from circular to off circular (elliptical), as much as 3% between our closest point to the sun and our farthest point from the sun. This orbit is influenced by our nearest neighbors, Mars and Venus. Eccentricity seems to have a major effect only when combined with the extremes of our precessional cycle and our axial tilt; the three cycles converge approximately every 100,000 years, and was first pointed out by Milutin Milankovich in the early 1900s. Therefore, the theory is called the Milankovich Cycle, and appears to be a factor in the coming and going of glaciation or Ice Ages, when other cataclysmic forces are not involved.

Since most of the Earth's landmasses are located in the Northern Hemisphere, the effect of our being farthest from the sun in our orbit combined with being farthest from the sun in our cycle of precessional can cause much more snow to accumulate on land. Our present elliptical position is at neither extreme, and only slightly off center. A 3% variation in our closeness to the sun can result in a 6% increase in insolation (solar radiation), but only a slight increase is predicted for the next few years. We have a tendency to 'forget' that the crust, 'our part' of this planet is only a 'line' thick when viewing the entire sphere – considering the Himalayas and the deepest ocean troughs, only 'thickens' the line slightly. There is this whole complex 'body' that has its core, its varying layers of magnetism, its tectonic plates of which we are mostly unaware. We worry about global warming, the ozone layer and the melting of the ice caps. But have we considered that Mother Earth is around four billion years old and seems to have a perfect 'balancing' mechanism in place? She may need to 'shift' some of the weight off the poles to accommodate for the increasing cosmic radiation causing magnetic changes, the pull of different forces as we line up with the center of the galaxy, and other factors we may not have a clue about!

So to bring it all together: NOAA reports that precession is in the 'cooling mode,' while axial tilt and eccentricity are in a slight warming mode. We seem to be "safe" on those counts. The Mayans may have given us a clue to their concern by making the 'womb' (center) of the galaxy their main interest. They knew of precessional (our changing orientation to the sky on the vernal equinox), but it is interesting that they named 2012 as the end of this 25,000 to 26,000 year cycle. LaViolette's theory that this coincides with the immense periodic explosions from the center of our galaxy, causing a major increase in cosmic ray impact on our atmosphere, adds a bit more concern to our scenario. Coming out of the galactic 'equator' plane makes the Earth more vulnerable to space debris, asteroids, and that interstellar 'dust' cloud, which could cause the thermal effect of heating our atmosphere. Then there is that asteroid belt between Mars and Jupiter as well as the Oort Cloud of asteroids and planetoids out on that leading edge ploughing through the galaxy. It all sounds a bit foreboding unless we take one other very large factor into consideration – consciousness.

Effects on Humans

We have not examined how all these 'sky' factors impact you and me, except in a 'whole Earth' scenario. "When early humans connected astronomical patterns with patterns of human behavior, they recognized a link between Earth's cycles and human physiology."[26] So states Dr. Bruce Lipton and I agree. "As above, so below." As he says, "If bacteria can evolve purposefully, then why not us?" Carlos Barrios calls astrology the science that focuses on the relationship between the stars, our planet and its inhabitants.[27] It has been practiced by all great civilizations. There was a time which we called the Dark Ages that the knowledge of this relationship had to be hidden and secret societies formed to protect and pass on this information. The twenty Mayan day signs were (and still are) used not only to discern a newborn's life purpose, abilities, gifts, and life cycles, but to help the shaman know when the power of a certain place was strong, when ceremonies must be performed, how to handle energy and predict the future.

The practice of following the Mayan signs enabled . . . a harmonious civilization, writes Barrios. No one lived in opposition to their own energies and destiny. Everyone respected each other because they knew they were all part of a . . . whole, part of the cosmic plan.[28] Their day signs had names like *Alligator, Wind, Hearth, Lizard, Serpent, Death, Water, Road, Jaguar,* and *Eagle.* "Hello, I'm a serpent; what's your sign?" It doesn't come across quite the same as our trendy sun signs of the Zodiac. But, that is not the type of astrology we are talking about here. These day signs each have a unique quality that not only influenced the character of those born on that day, but made performing certain tasks favorable or unfavorable. Some Mayans living in the highlands of Guatemala have been using these day signs in an unbroken sequence for thousands of years.[29] It reminded them that they were an essential part of the web of life – Earth, sky, plants and animals. Gregory Sams stated, "It was, perhaps, the intuitively arising belief that astronomical events had an influence upon our lives that led our distant ancestors to develop the science of astrology."[30] "As above, so below."

Most scientists, of course, discount the study of astrology as superstitious, perhaps because of the fact that it has been distorted from its original purpose by the popular horoscope. But its very persistence makes us want to look at it more closely. The influence of the sun was noted in the early part of the 20th century by a Russian professor of Astronomy and Biological Sciences, A. L. Tchijevsky, (also spelled Chizhevsky) in the connection of solar cycles (sunspots) and human moods, behavior patterns and creativity. The five years at the peak of each cycle, when there is greatest amount of energy and radiation, affect 'human excitability," leading to greater creativity and scientific discoveries or, according to our choices, more crime and more severe war incidents, as Tchijevsky knew from experience and observation during WWI.[31] Choice is the critical word here.

Gilbert and Cotterell applied this same approach to their study of the effects of fluctuations in the geomagnetic field on the human endocrine system, linking the periodic magnetic shifts or reversals in the sun to hormone production in the pituitary gland.[32] But there must be more to the relationship between sky and human that makes up the science of astrology. I think it has to do with a long history of humans seeing the cyclic connections between where the various parts of the universe were when certain events happened on Earth. We know that every bit of matter in this three dimensional universe that we know of is made of atoms which are mostly empty space and smaller 'entities,' for lack of a better word, that can switch from particle to 'wave' or energy and back again quite easily. These tiny 'bits' seem to have 'intelligence,' and a set of 'rules' that they go by. What we are building here is that everything from planets to quarks have a 'frequency,' energy, or vibration. We know that there is a 'something' that connects everything in the universe, and that the influence of magnetism is especially strong within our solar system and galaxy. Which bring us to how different alignments of sun, moon and planets can exert different 'force fields' upon the Earth and especially its life forms.

Cosmic Influences

*L*yn Birkbeck combines the observations of the past with quantum physics to show that we can create the future we want from whatever alignment of stars may come; he calls it quantum astrology (Astro-Quantum).[33] He sees the influences like a 'Cosmic Clock.' When two planets are at specific angles to one another, this period will have a certain quality, 'mood,' feeling, energy or spirit. We can look back at events in our history and discover how these 'moods' were handled by the mass consciousness at the time. Birkbeck calls these time periods 'waves' or "Planet-Waves." These waves can occur simultaneously, weaving a kind of 'time tapestry' of supportive or conflicting influences or a combination of these two. He sees these waves like weather; some can be stormy while others are generally favourable (UK spelling) and help us find solutions and create harmony. We can 'surf' these waves in an informed manner, taking advantage of the 'favourable times', or go out like a first time surfer into big breakers![34] These waves are a very suitable symbol for this inquiry since we saw above that everything is vibration. Thoughts are vibration or 'waves.' Light and sound are waves. Quantum physics has taught us that doing or thinking something in one place has an immediate and corresponding effect in other places and thereby affects the whole, and this applies to how one is aligned to the energies of the current Planet-Wave. Birkbeck says, "Planet-Waves are expressions of the Source before it has manifested as any event or human experience."[35] But both astrology and quantum physics are essentially spiritual in their outlook; as it is understood that life energy streams eternally from the Source and takes form as living beings . . . ad infinitum, Birkbeck continues. "Quantum theory can be seen as a metaphor for mind-power." " . . . the centre of the centre of the Uranus/

Pluto Mother-Wave coincides with the end of the Mayan calendar, 2012, which is predicted as the End of Time as we know it. As I see it, these various conjunctions have both a positive and a negative effect, such as his example of 1987-91, a Saturn/Neptune wave indicating 'dissolution of boundaries and disillusion with authority.' The Berlin Wall came down and the Soviet Union ended, but there were also 'boundary-breaking' incidents such as the Exxon Valdez oil spill and a massive earthquake in San Francisco Bay area.

"This Uranus/Pluto Mother-Wave 2007-2020 has the potential to boost human mind-power and enable us to make 'course corrections' that will avert disasters that threaten us."[36] Birkbeck relates that the character of Uranus is Freedom and Innovation and its energy is Awakening and/or Disrupting. The character of Pluto is Depths and Power and its energy is Empowering and/or Destroying. This has the combined effect of Awakening and/or Disrupting Depths and Power, and Empowering and/or Destroying Freedom and Innovation. The only other major storm of this time period is a Saturn/Uranus wave, which will be mostly behind us in 2012. The Saturn/Uranus combination in the past has been notable for its severe earthquakes and we have already seen several of these in this period. These inclinations are heightened by a full or new moon or any eclipses. In the end what it all comes down to, and I heartily agree with Birkbeck on this, is that **everything is interconnected and our state of consciousness is what determines our fate.** I also like his quote by Martin Luther King, Jr. "We must learn to live together as brothers or perish together as fools." Noting the present state of 'affairs' in the world on many levels, it appears only a miracle would help, but that term is relative – it all depends on what you believe is possible and I declare, "Let's set our 'sights' HIGH!" "You must Be the change you wish to see in the world." (Gandhi 1869 - 1948)

In *Toward 2012: Perspectives on the Next Age*,[37] Daniel Pinchbeck introduces us to the work of Richard Tarnas and his *Cosmos and Psyche*. Tarnas concurs with Carl Jung, who wrote, "Our psyche is set up in accord with the structure of the universe, and what happens in the macrocosm likewise happens in the infinitesimal and most subjective reaches of the psyche." He focuses on planetary transits – geometric relationships between the bodies of the solar system – and the correspondence that these alignments have with the dynamics of civilization. ". . . it is based .

. . on a deeper realization that human consciousness is meshed within the larger universe, a fractal that . . . expresses the larger pattern of the whole." Historical events back up his explanation of the conjunctions. In his words "When Uranus and Pluto come together, the party starts – and then tends to get out of hand . . ." "(it) amps up . . . the urge toward liberation and creative breakthrough, while incit(ing) rampages that often end in violence." The last such alignment came in the volatile '60s. Pinchbeck relates that Tarnas calls, not for fatalism, but for viewing the human condition as one of 'creative participation in a living cosmos of unfolding meaning and purpose.'[38] I especially like his integration of modern rational thought with ancient metaphysical principles. Pinchbeck states that Tarnas has given us a transformative matrix for reconceiving our relationship to the cosmos.

"Astrologers start from the assumption that planetary alignments have significance," writes Lawrence E. Joseph.[39] I agree with his assessment that a competent reading of one's chart can reveal past and future events, as well as hidden present conditions, to a notable degree. But . . . there is genuine scientific value to the study of planetary configurations. A cadre of space scientists now believe that planets exercise significant electromagnetic and gravitational influence on the Sun. It seems the Sun-Earth connection is a two-way street; that an energetic feedback system exists between them. Joseph continues, "Hurricanes, volcanoes, earthquakes, and other climatic/seismic events . . . could both cause and be caused by sunspots. Mayan shamans acknowledge the Earth's influence on the Sun and have done so for millennia."[40]

Space scientists such as Richard Michael Pasichnyk, who wrote *The Vital Vastness*, affirm that some configurations, such as lined up, opposite (180°) or squared (90°), may create rifts in the Sun's outer layer. According to Pasichnyk, "The Earth's magnetic field undergoes changes of intensity that reflect the magnitude of changes in solar activity *before* they take place on the Sun." It is interesting to note that the period of great hurricanes in 2005, bracketed the record solar activity of September 7 – 13; Katrina preceded it; Rita, Stan and Wilma followed it.[41] Joseph goes on to say that quantum physicist, Thomas Burgess, explains that the solar system's center of mass is constantly shifting and may be located as far as a million miles away from the Sun! The Sun wobbles and bulges in the direction of the center of mass, causing the possibility of a fissure, suddenly releasing what is known as 'imprisoned radiation.' Under normal circumstances this radiation leaks out in a more or less steady stream, but can erupt in a major outburst. Burgess also reveals that the Sun's magnetic poles reverse every 22 years or every other sunspot peak, and is due to reverse in 2012 adding more volatility to the situation.[42] So, we see that any further planetary transits may be dangerous.

In an interview Joseph had with Alexey Dmitriev, the geophysicist from Akademgorodok, Russia, who has written so much about the interstellar energy cloud we are approaching, made it very clear that he believes there are three significant factors that have not been emphasized. First, the condition in the medium of interstellar space (the *ether* comes to mind or plasma that we have discussed*) is not cold nor is it a vacuum-like void, but can be turbulent with electromagnetic activity. He says our solar system is moving into an area of 'magnetized strips and striations containing hydrogen, helium, hydroxyl (one H short of water)* and other elements and compounds, we normally call space debris.'[43] The shock wave on the 'bow' of our 'vessel' (solar system)* has become perhaps ten times denser and thicker due to this debris, which has led to a plasma overdraft and breakthrough into the region of the sun and planets. This causes the Sun to go into 'erratic behavior' and the Earth's magnetic field to come under stress. Part of his information comes from the cumulative data of Voyager satellites, NASA and ESA. The planets on the outer edge of the system are taking the worst of the 'shock,' especially Jupiter that has seen a doubling of its magnetic field. In 2006 our largest neighbor developed a new red spot, an electromagnetic storm almost as large as the Earth. Both Dr. Dmitriev and Dr. Vladimir Baranov point to the shock wave as the cause of these anomalies.

The second concern Dmitriev had, were the 'energetic effects of the planetary configuration.' Does this lining up of planets sound like astrology to you? The third, being 'the energetic impulses from the center of the galaxy,' brings us back to the 'womb' the Mayans spoke of, and that LaViolette thinks is due to emit a galactic superwave. Can *Gaia* adjust and regulate itself to compensate for these external disturbances?[44] That is the question! I am going to go out on a limb and postulate that it would take a LOT to completely overwhelm the perfection of this blue planet!

Unified Theory or Dimensional Shift

One final perspective before we move on: Barbara Marciniak, channeling the Pleiadians wrote, "At the precise time of your birth, the stars, planets, moon and sun were in a specific configuration. When you emerged from your mother's womb, the energy from the stars and planets imprinted your flesh... Within that moment were written certain probabilities, specific opportunities, and distinct challenges... Everything is in geometric relationship to everything else, creating energetic patterns. You yourself selected a moment and a lineage, a bloodline..., to give you the opportunities that *you** assessed would be ideal for you to experience in this lifetime. You determined these experiences according to what you needed to learn..."[45]

Ms. Marciniak also reveals via the Pleiadians, "The Mayans... have left you a number of clues...; their calendar... marks a time of ending and closure at the winter solstice in December 2012. The concept of the end of your time means that a cycle is coming to completion. It does not mean Earth is going to be finished...." "When that time is completed, there will be a *dimensional** shift upon this planet. Those who are able to accommodate the dimensional shift now are already moving into and out of... the fifth dimension."[46]

As complex as our modern science is with quantum physics, particle accelerators, chaos theory and more, it is still just dealing with the physical world of three spatial dimensions plus the dimension of time. Einstein and other scientists have spoken of other dimensions in trying to link the theory of general relativity and electromagnetic theory. Superstring theorists found the work of Polish mathematician, Theodor Kaluza, useful in that it postulated that space consisted of the three spatial dimensions we are familiar with and curled up circles within every point of our three dimensional space. Extend these circles to what they call Calabi-Yau (for the mathematicians at Harvard who described these) six-dimensional geometrical shapes, add our known three and time, and we have 10-dimensions.[47] (Yes, it is way beyond my brain's ability to conceive! Even a diagram did not help!)

The theory called *Standard Model*, based on bizarre particles such as quarks, leptons, and bosons, still did not pull everything together into a unified theory that links all matter and all forces. Physicists assume that during the Big Bang, six of the ten dimensions compactified into a tiny ball, while the remaining four expanded. Their problem is how this happened, leading them on endless trails of patterns of quarks, electrons, etc.[48] This has led to M-theory in 11 dimensions. Witten (who coined the term M-theory) and Paul Townsend (no relation to me that I know of) of the University of Cambridge has shown that these superstrings (which are basically vibrating resonances!) originate in the 11th dimension. The mathematics to prove this has been "fiendishly difficult – too hard for anyone to solve completely."[49] In 1997, Dr. Cumrun Vafa, Donner Professor of Science at Harvard, developed F-Theory, proposing there may be a 12th dimension out there, after years of research in the nature of quantum gravity and superstring theory.

But, aren't we still talking about the physical universe? So perhaps "another physical dimension" is not what we are working toward examining. Other terms used are parallel universes, alternate universes or other planes of existence when referring to 'higher' spatial or non-spatial 'places.' Perhaps we are trying to find that interface where non-physical becomes physical, the Source of everything, or even that place we call 'heaven.' An Infinite Intelligence permeates all dimensions including space/time and our bodies, and connects them intimately to our mind/souls/consciousness. American astronaut, Edgar Mitchell, said, "The consciousness of man has an extended nature which enables him to surpass the ordinary bounds of space and time – suggesting that there is another dimension beyond the material world."[50]

In *The Key* Whitley Strieber said The Master defined the soul as a "radiant body, formed out of conscious energy" . . . "part of the electromagnetic spectrum, easily detectable by your science."[51] That sounds very much a part of the physical world, even if mostly invisible to our current 'perception.'

SIGNS AND PROPHECIES

Predictions

Reviewing James Redfield's *World Vision* in his third book, *The Celestine Vision*[1] published in 1997, I am sad to report that as of 2011 not a huge amount of his vision has materialized. One aspect of his vision that does seem to be happening more and more is the contact many people are having with other 'dimensions.' There are more 'out of body' experiences (OBEs), more channeling of discarnate entities, more near-death experiences with life-changing aftereffects, more visions, spiritual experiences and the remembering of past lives. As I write this I am looking forward to reading this visionary's latest book *The Twelfth Insight*.

Ken Carey was another author from the '70s and '80s who spoke of the Mayan 'end time;' he said his information was received in a direct transmission, not channeled, and, we are assuming this means from another dimension or, as he said, an angelic source. He spoke of the need for a critical mass of the population to be on a 'higher' frequency, or more spiritually evolved as the Transcendental Meditation movement advocated. Geoff Stray quotes Carey: "There will be a great shift then, a single moment of Quantum Awakening . . . the smallest interval of time . . . will be lengthened unto infinity. Through that expansion, eternity will flow. In the expanse of the non-time interval, human beings will have all the time they require to realize, experience, and remember . . . their eternal spirits . . . Each one will have the choice to return to biological form or to remain in the fields of disincarnate awareness."[2] (Perhaps this is the end of 'time' as we know it.

In 1975 Dannion Brinkley was struck by a bolt of lightning through his telephone line and, basically, died. He revived 28 minutes later and came back with an incredible story. He said he was shown over 300 scenes from the future and was told that one hundred and seventeen of these scenes about major world events would take place before 2012. There seems to be some evidence that some of those events that he spoke about have happened and, from the list of those yet to happen, some seem very likely, such as oil being used as a weapon to control world economy. As I attempted to verify his predictions I found an interesting web site on near death experiences: www.near-death.com/experiences/research32 which is not only a very complete listing but gives much of the spiritual advice brought back by those who encountered 'something beyond the veil.' I liked the section entitled "A Successful Apocalyptic Prophecy is One That Doesn't Happen." As Kevin Williams, the site originator (?), says, "A better understanding of prophecy reveals that either (1) or (2) is true: (1) The prophecy was successful in permanently diverting the outcome by the raising of the world's consciousness, or (2) Because the prophecy gives an exact date, the prophecy may still be valid and the date may be wrong.

"Dr. Kenneth Ring, who completed a huge study of NDEs called The Omega Project, says these permanent changes (after NDE) consist of improved self-esteem, more altruism, a less materialistic approach to life, a more broad-minded and tolerant approach to spiritual and religious beliefs, and many instances of telepathy, clairvoyance, precognition, and other psychic abilities." Pam Reynolds returned from a NDE predicting Earth and consciousness changes as well as a physics breakthrough in 2012. The physicist, Alexey Dmitriev believes that the increasing plasma balls in the atmosphere indicate an imminent "transformation . . . involving an interaction with processes beyond the three-dimensional world."[3] (Geoff Stray)

Much, it seems, depends on free will, our choices. I also found that a web site called nderf.org that has a listing of thousands of people who claim to have had near-death-experiences, many more than I expected! Many speak of the time around 2012 as significant. Brinkley said (in summation) it was his hope that as the Earth changes happen, we all seize the opportunity for growth. Louis Famoso said after 'coming back' that he was told, "Look to Orion for a sign."[4] (Cryptic!) Solara writes in her book, *11:11*, "The three stars in the belt of Orion . . . are the main control points, or pins, which hold our dimensional* universe into position."⁵ Does this bring to mind the three hearthstones of the Maya that represented these same stars? Mellen-Thomas Benedict, another with an NDE, said, "We are multi-dimensional beings. We can access that through lucid dreaming.⁶

Dimensions

The dimensions that spiritual guides speak of appear to have a different 'definition' from the dimensions that the superstring theorists have in mind. Stephen Hawking, famous theoretical physicist, speaks of eleven dimensions of nature. But do all eleven of these levels fall under the same 'laws' as those which govern our physical world? We think of the four dimensions that we are aware of as our physical reality. What we label the 'higher dimensions' we usually identify as the 'non-physical.' Are the other seven 'supernatural,' and outside the 'laws' of physics and quantum physics? Carlos Barrios gives us a clue to other 'realities': "The Grandfathers call these unreal dimensions "unconfigured space." The *Popol Vuh* speaks of parallel dimensions as paradise and calls them *Paxil* and *Kayala*. . . It's like going back to *K'ajolom*, the matrix, the origin. *Njat* (space/time) is manifested differently. . . . Light behaves differently in these parallel dimensions . . . we are accepted unconditionally there – we are truly loved."[7] Sounds like what we call the afterlife! Barrios also writes, "The Maya were obsessed with breaking through these borders (of what we call reality)* to travel (beyond) space-time." "The idea is to move forward and exceed the speed of light. If we are able to manipulate the light, it means that we're able to shatter constraints inside our spirit and consciousness and go beyond it."[8]

Another reference of this type of 'dimension' is from Sheila Gillette who channels THEO, a group of twelve Archangelic beings, or teachers, who disclose, "We come, vibrationally, into the sixth dimensionality arena to communicate. However, there are twelve dimensions about the earth and we are of the higher dimensional realm. . . For there is a great shift and change energetically at this time. . ." "Consciousness of humankind (is) raising to a higher vibrational frequency and we are facilitators, mentors. . . ." She goes on to conclude that THEO has stated that the year 2012 will be the beginning of a thousand years of peace, and that the evolution of humanity will not reverse itself.[9]

Barbara Hand Clow, in her 'transmissions' from the Pleiades via *Satya*, describes Earth as having nine dimensions of consciousness that all humans can access. "This ability, which was natural for humans long ago, is opening up now . . . (and) is accelerating as we approach the end of the Mayan calendar, December 21, 2012."[10] She describes these dimensions in *The Pleiadian Agenda*, and in *Alchemy of Nine Dimensions*, she goes into the scientific explanations of and comparisons to Carl Johan Calleman's evolutionary time acceleration theory and its connection to the Mayan calendar. According to these writers, during the final and ninth Underworld, time will speed up by twenty times in February 2011.[11] This is a rather frightening prediction, since it is February 26, 2011 as I write these words! David Ian Cowan further illuminates these ideas about dimensions, time, duality and electromagnetic energy that were first presented by these two authors, in his book, *Navigating the Collapse of Time: A Peaceful Path through the End of Illusions*.[12] His comparison of the Mayan Tree of Life or World Tree to the collapse of Time is very thought-provoking.

Other cultures, such as Tibetan Buddhists, ancient Egyptian, and the mystical side of Judaism, Kabbalah, have all referred to these realms or 'levels' by various names.[13] The Mayan shamans have, at times, accessed these 'dimensions' with hallucinogens such as peyote, ayahuasca, and the *Bufo alvarius*[14] toad potion. A number of people today are accessing other 'realities' without any help from 'substances.'

Gregg Braden describes dimensionality as a shift in vibration. ". . . an attempt within the logical mind to resolve the heirarchy of energy into discrete bands, or ranges of experience, so that they may be referenced separately. The reality is that there is no separation. There are no clear boundaries between one vibrational experience and another. There is only . . . a phasing of varying states of experience. Dimensionality may be considered from the perspective of one substantive energy expressing itself differently in response to changing conditions, those within which frequency may increase or decrease. . . . All that exists within creation . . . is . . . patterns of waves interacting with waves. The indigenous peoples . .

. have told us for thousands of years that our world is the world of illusion; that all in this world is the product of thought and as such is not real."[14]

'A vibrational level of experience,' a 'zone' of information, a shifting of resonance are all ways of describing 'dimensions,' but are we any closer to really understanding what that means? Is it pertinent to our inquiry? I believe it is if we are going to comprehend what the Mayans were attempting to relay about this end date. "The shaman . . . is occupying, in a perfectly alert way, an extended reality that includes the worlds of matter, energy, power and spirit."[16] F. David Peat goes on to state "that shamans and Native healers are judged psychotic according to Western psychological tests. Could it be that our own society, with its rigid, compulsive grip upon a single reality, is the one that has become abnormal?" Remember, the word *samân* is found in the Tungusian (Where have we heard that word before? Oh yes, as in the meteorite that exploded above Tunguska, Siberia, flattening hundreds of square miles!) language spoken by Indigenous peoples of northern Siberia. He also relates that not all healers, especially Native American, wish to be called 'shaman.' "Shamanism, with its emphasis on trances, visions, and altered states of consciousness, has become synonymous with Native American spirituality and Indigenous science . . . which some non-natives believe can be learned in weekend workshops,"[17] Peat writes.

"For thousands of years the shaman has deliberately set about to suspend, at will, the confines of the ordinary worldview and, through experiencing that ecstatic state of oneness, has been able to bring back one more piece of the puzzle of the true nature of our cosmos," says Nan Moss in *Weather Shamanism*.[18] So is this 'extended reality' part of the 'other 10 or 12 dimensions' physicists speak of, or is it something entirely different? When someone receives information 'from the other side,' when someone talks with their dead relative, when a person has an 'out of body experience,' when a crop circle 'appears' in a matter of seconds, or when 'orbs' show up on photographs, especially around spiritual gatherings and healings, what 'other reality' do they come out of? Are most of us, even our scientists, just not able to detect these other dimensions? Some call these types of anomalies 'primitive,' superstitious, or even pagan, and others say this dividing line, this 'veil,' is getting 'thinner' or 'lifting' as we get nearer 2012. Or is consciousness evolving to a point of not needing this reality or dimension blocked? I suspect that 'door' has always been 'unlocked' or even open from the other side! If this veil lifts or tears too quickly, will humankind be able to handle this more expanded view? Or will we be like some primitive cultures when exposed to technological civilization – have a sense of futility, irrelevance, and disempowerment?

Lipton and Bhaerman have a very interesting chapter in their book, *Spontaneous Evolution*, which speaks about this 'line' between Spirit and Matter. They explain with a graph that is half Spirit (in white) and half Matter (which they show in black) with a horizontal line between them, then blending the two in shades of gray from white to black. They begin by placing Animism (8000 BCE), believing that spirit existed in all things – animate or inanimate – on the center line, "a perfect balance between the spiritual and material realms." They continue on to place polytheism (2,000 BCE) higher in the Spirit range and monotheism (800 AD) higher still. But with the Reformation (1500 AD) the line begins to curve downward and drops to the line again with Deism (1776). Darwinism (1859) and Neo-Darwinism (1953) drop way down into the darker gray/black part of the graph. The authors postulate that the Human Genome Project (2001) which, in identifying only 23,000 human genes, glaringly revealed that "the neo-Darwinian belief in a genetically programmed biology (was) fundamentally flawed."[19] This brought the line halfway back to mid point or balance. "Just as animists and deists understood that spirit and matter must fully co-exist,* we are being challenged to move past 'either-or' and to recognize 'both-and.'" The final graph shows "Holism" (2012) at the mid, balanced, line again. Then they ask, "Is it possible that by integrating the opposites of spirit and matter, energy and particle, masculine and feminine, we can create an emergent human society, one never-before-seen, whose expression is completely unpredictable by studying what we have and who we are now?"[20]

Hazel Courteney in her 2010 book, *Countdown to Coherence*, quotes Dr. William A. Tiller's illuminating explanation of these dimensions or levels from his *Conscious Acts of Creation*. He divides physical reality into two distinct levels: the 'dense' electric atom/molecule physical world and the "normally invisible magnetic information-wave level," sometimes known as the etheric level, which "travels faster than the speed of light and is outside of space-time, but can be influenced (by) consciousness." Tiller's experiments show that the magnetic information wave level is the template upon which the atom/molecule level is built, but that these two "levels remain 'uncoupled' most of the time. The two levels are 'coupled' by human consciousness and focused intention. [21]

Dr. Peat mentioned above, who is a theoretical physicist and has worked with the famous physicist, David Bohm, is also completely knowledgeable about the Native American worldview. He points out that Quantum theory stresses the irreducible link between observer and observed and the basic holism of all phenomena; Indigenous science also holds there is no separation between matter and spirit, between each of us and the whole of nature. Physicists know that the essential stuff of the universe cannot be simply reduced to atoms, but exists as relationships and fluctuations at the boundary of what we call matter and energy; Indigenous science teaches all that exists is an expression of relationships and balances between what we would call energies, powers and spirits. Flux and process are fundamental to both modern and Indigenous science. In line with some leading edge physicians, Native healers suggest that healing involves the whole person – body, mind, and spirit. Indigenous as well as modern ecologists stress the interconnectedness and wholeness in nature and (that it is) inherent within all of life.[22] So, let's explore some of these phenomena that seem to breach this boundary between physical and non-physical. I hope you are prepared to journey 'out there on the edge!'

Orbs

Since the digital camera came into popular use, photographers all over the world have been reporting unusual circular-shaped optical anomalies on their photographs that were not visible when the picture was taken. After trying to explain them as atmospheric pollution or problems with the camera, the skeptics could not contend with the enormous number and specific alignment of this type of 'occurrence.' The fact that William A. Tiller, Ph.D. (Professor Emeritus of Stanford University) wrote the Foreword to *The Orb Project* by Míceál Ledwith, D.D., LL.D. and Klaus Heinemann, Ph.D. gives credence to their work and his extensive explanation gives even more. Dr. Tiller starts by saying that "(orbs are) a part of the heightening of awareness brought about partially by the elevation in human thinking and partially by the increase in energies directed toward this planet by . . . life-forms existing in . . . (unseen) dimensions." His field of expertise is 'psychoenergetic science' which expands the $E=MC^2$ to Matter = Energy = Information = Consciousness. Research in the field of consciousness, especially with placebos, is nothing short of amazing. Dr. Tiller worked with the Institute of Heart Math (We will talk more about their research in the next chapter) measuring the heart rate variability when the 'heart' (emotion of appreciation) interacts strongly with the (visualization) brain, showing a strong positive correlation and a significant change in chemical output such as DHEA (the precursor to most hormones). In the '70s he worked with Stanislav O'Jack who always seemed to have a lot of strange phenomena, including orbs, appear on his developed film, more when he carried the camera next to his body for several days. After experimentation, Dr. Tiller found that the man's biofield sensitized the camera to access another level of reality. Research has shown that a space can be conditioned by specific intentions to a higher electromagnetic gauge symmetry state than that of our normal level of physical reality, in other words, "human consciousness is capable of coupling humans and instruments to another unique level of physical reality, not normally detectable by conventional instrumentation . . ."[23] We frequently call this level – sacred space.

Dr. Tiller explains these anomalies in greater detail in an interview with Hazel Courteney in 2009 (in *Countdown to Coherence)*, ". . . the camera traditionally uses photons of electromagnetic energy. But there are a host of other energies that go faster than such light, which can pass right through 'solid' objects – radio waves, for instance. (We) have known for years that we and our world, apart from minuscule amounts of matter that we can see, mainly consist of 'empty space', but this vacuum level of physical reality is not 'empty' at all. The energy stored in all the physical mass of our universe is minuscule in comparison to the energy stored within the 'stuff' of this vacuum. It is (the) highly intelligent latent potential that underlies all of life, and what we need to do is to 'couple' with it through intention. . ." Tiller also calls this vacuum - the magnetic information wave level.[24] Others have called it the 'etheric' level, the Matrix, the Field, 'the soup of potential', Infinite Mind, the unseen dimension, the infinite web of connection, or the implicate order (Bohm).

"Orbs cast a new light on the philosophical/spiritual/scientific realization that "All is One." "The material (in *The Orb Project*) is the result of several years of careful, entirely independent research of the orb phenomenon by two authors who had not known each other and who were unaware of each other's work prior to the publisher's initiative for (the) collaboration."[25] Dr. Heinemann compares the orb phenomenon to the apparitions of Medjugorje (40 children and young adults saw and conversed with the Holy Mother [Catholic]) in ". . . that highly evolved Spirit beings in the Spiritual reality see that humanity is in dire need of discovering its purpose and that they devised phenomenal occurrences . . . to help mankind in this epochal jump in consciousness."

From their research the two researchers have found (1) an orb is a *physical* manifestation . . . (that) has a mechanism to manifest itself . . . as electromagnetic waves in the visible spectral range, and (2) the orb's mobility must be assumed to be entirely directed and controlled by the Spirit being from which it emanates. Dr. Heinemann concludes that these

Spirit beings are (1) all around us, (2) highly evolved and intelligent, and (3) capable of changing their size and location extremely quickly. "We can summarize these conclusions in one important statement: Photographs of Spirit emanations offer evidence – as close to scientific proof as we have ever come in proving the existence of Spiritual reality – that Divine Presence is real."[26] To find the details of their research, read their very interesting book. Dr. Heinemann has written, with his wife, Gundi, a follow-up book with many other photos of orbs.[27] They quote a Canadian man who took photos at the Casa de Dom Inácio during a healing by Medium João, "Orbs are 'doing' something that we cannot see with other instruments. We do see them at one particular moment of their existence, 'frozen' for a fraction of a moment in a four-dimensional space-time position."[28] I also found this subject impartially researched by Jane S. McCarthy of South Wales, UK on her web site: http://www.psychicinvestigators.net/html/orbs.html.

Crop Circles

I consider crop circles another 'sign' that we are receiving some 'special' clues from 'higher intelligences.' No, I did not say extra-terrestrials; though some of those may be around (and a few people want to credit them with the circles); I just have not seen them! I have credible friends who say ETs are 'around'. My question is where? In one of these parallel universes or dimensions? The figures, no longer called circles, but formations, have grown more and more complex over the years. There is a report that they were seen as far back at 1678 in Herefordshire, England. Some appeared in Tully, Australia in the mid-60s. From then until the 1970s those in the UK were spoken of as 'UFO nests.' Geoff Stray reports that they were thought to be weather phenomena in the '70s and '80s, but soon became so complex, that explanation would no longer 'fly.' Hoaxers have tried to simulate the figures, debunk them and make fun of them, but yet the phenomenon continues. 1990 brought long pictograms of connected glyphs.[29] That year alone had almost 800 reported crop formations. (See Illustration #8 used by permission of Steve Alexander www.temporarytemples.co.uk)

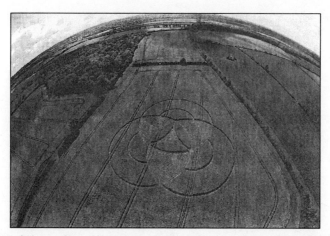

Trefoil crop formation, Hannington, Wiltshire, UK, May 7, 2011

Freddy Silva, who lives in Wessex, southern England where 90% of the circles have 'appeared,' has studied the phenomenon comprehensively and written the book, *Secrets in the Fields*. As he says, "the science (at the heart of the crop circle mystery) is so subtle, so wise, and so awe-inspiring that it has the power to humble you and wake you up to a greater reality." [30] The fact that many of these formations are near sacred sites, especially the megalithic shrines like Stonehenge and Avebury in England, leads us to wonder about their placement. What strange attraction is here? These sites as well as the 'circles' are frequently at crossing points of an invisible electromagnetic grid, geodetic energy lines, sometimes called 'ley' lines, that encircle the Earth. "At these terrestrial points, the veil between worlds is thin, and the concentration of energy . . . influences the rhythms of the human body . . ." Once again we notice that our planet's magnetic field is being pointed out. The perfect, intricate patterns demonstrating Universal principles of geometry (some mathematical theorems not known before!), and ancient religious symbols (some appear to be Mayan), to name only two categories, exhibit mathematical precision. One (Silbury, 1991) encoded 19.47^0, the latitude at which energy upwells on many of the planets in our solar system - another reference to our magnetic field. Others were named dolphinograms and insectograms, one near Stonehenge, its message unknown, was, perhaps, jest. The majority of formations usually appear from April to September in the wheat and grain fields, each stalk is bent an inch above the ground, the swirls include every stalk in perfect alignment, any blossoms, such as on canola, are undamaged, intact, and designs of surgical precision sometimes include thin rows left standing or thin swirls going the opposite direction; every new formation is unique. There have been lights (some laser or tube-like) sighted, and harmonic tones or humming heard, in the vicinity of formations in progress, but never any entrance or exit evidence in the 'real' ones. Animals have been known to react strangely around the circles and in Canada a flattened mummified porcupine was found in the center of one.[31]

These cryptic, graphic 'communications' have appeared all over the planet, from Japan, Australia, Afghanistan, Russia, Egypt, Canada, Hungary, Ireland, Belgium, New Zealand, Turkey, Brazil, Puerto Rico to South Africa and the U.S., in mediums as diverse as snow, ice, encrusted sand (Egypt) and a dry lake bed in Oregon. Formations in plants, up to 150,000 square feet large, everywhere exhibited the same features: undamaged plants, precisely swirled, soil underneath not compacted. Tests done on the plants from authentic 'circles' showed that an energy of some type had altered the plants' crystalline structure and heating of the interior had caused tiny holes at the node of the stem created by superheated, pressurized vapor forcing its way out, as was duplicated in a microwave oven; this is not found in normal crops under any known circumstances.

Silva quotes Dr. W. C. Levengood: "the affected plants have components which suggest the involvement of rapid air movement, ionization, electric fields, and transient high temperatures combined with an oxidizing atmosphere. One naturally occurring . . . force incorporating each of these features is an ion plasma vortex."[32] He goes on to explain further research: Microwaves are an electromagnetic energy wave with a frequency above 1GHz and quickly dissipate after initial contact, yet electromagnetic sensors show an energy residue remains in the circles, sometimes for years. So due to this and other effects, such as rendering seeds sterile, microwaves were ruled out. Japanese scientists have demonstrated that plasma (superheated ionized gas, which we talked about in the previous chapter) is capable of organizing itself into simple concentric shapes with alternating positive and negative charge as was seen in the early circles and rings, but cannot apply to the more complex glyph and pictographic formations. There is no doubt some intelligent force is behind this phenomenon since it avoids houses, towns, people, and gardens, and produces hundreds of philosophically significant shapes of great complexity. Visitor's photographs in and around the crop formations frequently come back with 'orbs' as we discussed above.

"When objects from other levels of reality alter their rate of spin (or frequency) they are *observed* as increasingly physical phenomena in our *dimension*,"* writes Silva. "Studies into bioplasmic energy fields and orgone energy . . . show that we are surrounded by a life-force energy which is interactive, yet seemingly invisible to the limited human range of vision (such as auras)." "That this force is associated with consciousness was hinted by Nobel Prize-winning father of quantum theory, Max Planck: "There is no matter as such! All matter originates and exists only by virtue of a force. We must assume behind this force the existence of a conscious and intelligent Mind. This Mind is the matrix of all matter."[33]

Crop circles have many times duplicated icons, petroglyphs, symbols and hieroglyphs from ancient and indigenous cultures, who see (or saw) these circles as receptacles of Divine truth, vessels of communication between two worlds. One symbol that expresses the Golden Ratio 1:1.618, the spiral, seen in many formations, has been, through the ages, a symbol of the manifesting of matter from non-matter (spirit). It is the shape of our galaxy, seen in the pine cone and sunflower, captured on the stone walls at Newgrange and, frequently, seen in Native American petroglyphs. (See Illustration #9 Spiral petroglyph)

Native American spiral petroglypth

The Native American medicine wheels remind us of the universality of the symbol of the circle for Oneness and Wholeness, for the cyclic nature of life and of our planet. Earth Mother with arms outspread is another ancient, well recognized symbol, as well as the moon in all its phases and the planets, all shown many times in crop formations. One formation mimicked the tiles inside the pyramid at Saqqara, Egypt, with the usual sacred geometrics. Everything from Sacred sound and Sanskrit chakras to Fractals and the Hindu Lotus have been displayed for our enlightenment. As Silva says, "The Circlemakers are using symbols based on Universal Principles."[34]

People in Hungary believed a healing energy existed in the 'crop markings' and took their sick children there to be cured. EEG tests show heightened right hemisphere activity in people inside crop circles, but not on stepping outside it. Believe it or not – people have had physical symptoms subside just by having a photo of a certain pictogram shown to them or placed on the affected body part. Sufferers of arthritis, osteoporosis, allergies, and tremors have been cured or temporarily helped by being in a crop circle. Modern medicine now knows that low levels of electromagnetic energy (ultrasonic) heals certain ailments, particularly bone disorders. This is "because the body is nothing more than molecules in a permanent state of vibration." Many people experience sharpened awareness, euphoria, calmness and joy inside 'circles.'[35] Onset of untimely menstrual periods has been reported as well as dehydration and nausea, especially if in the circle for a long time or if walking clockwise within it. A noted psychic said she was 'told' that the center of the crop circle was 'the center of all possibilities,' and that healing would be available from it. (Makes me want one in my backyard!) A person eating the seed from wheat inside the crop circle reportedly was cured of a malignant tumor. To say the least, Silva's book makes fascinating reading!

This, to me, is another example of an obvious 'crossing of the boundary' or 'veil' between worlds, like channeling, or healing by calling on what one may consider a divine being. Are these types of events happening more often? Yes, some phenomena have increased but the crop circle phenomenon is less active now than in the 1990s, at least in England, where they have been most prolific. One of the largest, over 900 feet in diameter made up to 409 circles of varying sizes in a spiral design, was found in Wiltshire.³⁶ As we saw earlier, the spiral has over the course of human history been an abiding symbol probably due to its repetition in nature, from the sea shell to the tornado. There has been some effort by the governments of the various countries to discredit the origin and meanings of the formations and discourage media attention, as with UFO reports and other "crossing of the dimensions" type of events. Perhaps, to "keep a lid on" any panic that they imagine might ensue! The same has happened with some discoveries in the field of astronomy and medicine.

Dr. Levengood (quoted above who is a biophysicist specializing in agriculture and seeds) formed a research team with Nancy Talbot and John Burke to study the crop circles. (BLT team) They found that the stems are softened so they can bend without breaking, the electrical charge on the stems is proportional to the amount of bending, and that the tissue around the seed has an increased electrical conductivity. Found in the soil of many circles was magnetite – meteoritic dust that has been heated and melted into microscopic spheres: a spiraling plasma tube would bring that down and concentrate it in the circle. An area such as Wiltshire, with an underlayer of chalk which is a natural aquifer, would attract the negative charge of the plasma. This brings us back to Dr. Dmitriev's theory of the magnetized plasma band that our solar system is travelling into at this time. But where do the exquisite and mathematically perfect designs come from? Geoff Stray reports some researchers suggest they come out of the 'global brain' or mass consciousness.³⁷ What about the formation that showed mathematically that there are three speeds of light, which seems to be unknown to our leading edge scientists?

Stray also reports on research into unexplained lights in northeastern Norway, which scientists believe are thermal plasmas that become visible when the geomagnetic field is perturbed (love that word the scientists use!) Stray, himself, experienced these lights with an infrared scope with Nancy Talbot of the BLT team who took flash photos over a crop circle in the Netherlands which showed numerous large and small orbs, though none were visibly seen at the site. Thus, there seems to be some evidence that the balls of light could actually be life-forms.

Further research collaborations have suggested that the air above a fault line (where layers of rock grind against one another) would be ionized (having atoms that had lost or gained one or more electrons) and could result in a plasma light orb. Dr. Dmitriev believes that ball lightning, earthquake lights, and plasmoids, are associated with *Zero Point energy in vacuum domains* (I won't try to explain, but let's just say that type of energy is HUGE!), as well as the geomagnetic field. The magnetic field of an area jumped 800% when the anomalous lights, known as *min-min light* in the Australian outback, were present.[38] None of the "scientific" explanations tell us how these impersonal forces manage to avoid all people, houses, towns, etc. and causes us to recall that quite a few people (including Terence McKenna) have predicted that "the laws of physics will be transformed around 2012."[39] The crop formations are like a lesson placed on the blackboard (Oops, I'm dating myself!) to be 'erased' at the end of the season, when the 'lesson' is, hopefully, realized. Will we be transformed? Let's look at some of the prophecies.

Ancient and Indigenous Prophecies

John Major Jenkins quotes Maya leader, Don Alejandro: "2012 was prophesied as a time of change and a new dawn, but the 'task to be finished' involves human . . . choice." "13 Baktun . . . is the time of . . . the return of the men of wisdom." ". . . rich or poor, black or white, men or women, indigenous or non indigenous, we are all equal, we all have dignity, we all deserve respect, we all deserve happiness; we are all useful and necessary to the growth of the country." "The world is transformed and we enter a period of understanding and harmonious coexistence. . . . The Mayan prophecies are announcing a time of change. The *Pop Wuj*, the Book of Counsel, tells us, 'It is time for dawn; let the dawn come, for the task to be finished.'" [40]

José María Chan, a Quiché day-keeper from Chichicastenango, spoke about 2012: "It is an event that has already begun. . . . Humans more than ever should pay close attention to all the events that disturb balance. . . . (In the) years previous to 2012 we have been experiencing different stages of a sacred effect that can turn harmful if we lose human wisdom. . . . 2012 is just the high point of the story."[41] Jenkins points to the problem of 'trying to fit the vast vision of ancient metaphysical wisdom into the small box of materialistic science.' "I believe that prophecy is best understood as an evocation, an ecstatic calling into being, of the highest possible outcome. This is the role of a true prophet." ". . . we should not pour our energy into envisioning (worst case scenarios) and projecting them into manifestation with our fears."[42] (Well said, Mr. Jenkins!) He also relates that 'apocalypse' is best understood in its original etymological sense: 'unveiled.'

Barbara Hand Clow writes, "According to indigenous traditions, the Light is infused with cosmic information, and modern science has discovered that photons carry cosmic information. Megalithic (huge standing stones)* astronomy, as well as (other types of)* indigenous astronomy, suggests that Light is more potent and transmutative for humans during the equinoxes, solstices, and new and full moons."[43] Or, as Gregg Braden says in *The Isaiah Effect*, "... ancient prophets foresaw equally viable futures of peace, cooperation, and great healing for the peoples of the earth ..." "... rare manuscripts ... left the secrets ... that allow us to transcend catastrophic prophecies and predictions and the great challenges of life with grace."[44] He goes on to assure that ancient and indigenous traditions, such as native North and South Americans, the Tibetans, and the Qumran (Essenes) communities of the Dead Sea, offer a framework that allows us to make sense of the chaos in our world. They provide a unified view of creation, reminding us that our bodies are made of the same materials as our earth. The Essenes, authors of the Dead Sea Scrolls give us the clear insight that "events observed in the world around us mirror the development of beliefs within us." (Can you say, 'as above, so below'?) The Essenes invite us to view ourselves as one with the earth, rather than separate from it. And as Braden says, "we are refining our choices in every moment."[45]

In the Hopi (a Native American tribe in AZ) worldview we are in the fourth (just like the Maya) of seven eras, and are now in the 'great purification' period before the start of the Fifth World (called the World of Illumination). The Hopi have a ceremony in November called *Wuwuchim*, held in an underground *kiva*, in which the initiates undergo a spiritual rebirth. As they emerge from the kiva, they are symbolizing the forthcoming emergence from the Fourth World into the Fifth World, just as the hole in the floor of the kiva (called the *sipápu*) represented their emergence from the Third World into the Fourth. This ceremony is at midnight when the Pleiades are overhead just like the New Fire Ceremony of the Mayans.[46] (Geoff Stray) The Hopi symbol for emergence is the labyrinth. They speak of five psychophysical centers similar to the chakras of Hindu and Buddhist traditions and believe that the Earth has similar vibratory centers along its axis in which each era is represented by a center, our present one is the most material and the next World starts back up the scale, less material.[47] Many Native Americans refer to drumming as 'the heartbeat of the earth.' The frequency of the earth is 7.5 cycles per second which scientists have discovered to be equivalent to brain waves of high theta/low alpha, which are related to creativity, vivid imagery, and states of ecstasy. Sandra Ingerman writes, "It appears that drumming allows shamans to align their brain waves with the pulse of the earth."[48]

The Hopi prophecy of the *Blue Star Kachina* is quite mysterious; *kachina* is their word for 'spirit,' or nature spirits, and they predict the Fifth World will begin when *Saquasohuh* dances in the plaza. This is their word for the Blue Star Kachina and there have been numerous conjectures about what this means. Frank Waters, who lived with the Hopi for three years, said it is a far off, invisible blue star. A Native American named White Feather said this ninth and last prophecy is that a blue star will fall with a great crash and the ceremonies of the native people will cease. Some say this has already happened, others say it is a comet, or Sirius C, which astronomers have recently discovered to be a 'red dwarf' star. The Dogon of West Africa have for centuries said the 'Dog Star' (Sirius) is a binary star system (discovered to be true in modern times) and based their calendar on the orbit time of Sirius B around Sirius A, saying that a third star existed. They have always said visitors (whom they call the *Nommo*) from this star gave them this information and would return one day.

The Puebloan people of the southwestern U.S., said to be descended from the ancient Anasazis of the Four Corners area, closely agree with the Mayans on the end date of this world age. One elder said the final cleansing began in June of 1998 and will end on December 22, 2012, and "If mankind will not willingly let go of the illusion, and the lie, it will be stripped away." The first day of the Fifth World (they say) will be December 23, 2012.[49]

In *The Divine Matrix*, Gregg Braden speaks about the unifying field of energy that connects everything through what we think of as 'empty space,' and compares it to the Buddhist Sutras which state, "in the heavenly abode of the great god *Indra*, there is a wonderful net which has been hung ... (and) that it stretches out infinitely in all directions." He then compares this to the Hopi creation story which says that "our world began long ago when Spider Grandmother emerged into the emptiness of this world. The first thing she did was to spin the great web that connects all things, and through it she created the place where her children would live."[50] Braden relates another story he heard from a Native American wisdom keeper, "A long time ago . . . there were fewer people and we lived close to the land. People knew the language of the rain, the crops, and the Great Creator. They even knew how to speak to the stars and the sky people. They were aware that life is sacred and comes from the marriage between Mother Earth and Father Sky . . . there was balance and people were happy."[51] Today, we have little other than the ruins and the petroglyphs (rock art) to give us clues about their world. (See Illustration #9, above) "The Hopi orientation bears no relation to North or South, but to the points on the horizon that mark the places of sunrise and sunset at the summer and winter solstices."[52] The story of Grandmother Spider brings to mind a crop circle near Avebury in Wiltshire, which very much looks like a spiderweb within a circle, which is said to encode a series of pentagrams with ratios that correspond to the production of sound. The Hopi legend that Spider Grandmother 'sang' into being the many forms of life on earth, makes the spiderweb representative of the Word of God. (From the "2010 Crop Circles" calendar by Freddy Silva, published by Amber Lotus Publishing, 2009)

The blog from www.temporarytemples.co.uk September 14, 2011 reviews Dr. John Stuart Reid's presentation at the Summer Crop Circle Lectures, who spoke about the relationship between vibration and shape and mentioned the German physicist and musician, Ernst Cladni, who first demonstrated this relationship. Dr. Reid said, "Simply put, shape is a fundamental response to vibration, and it is ultimately vibration that shapes all matter in our universe."[53]

There is a sandstone cliff on Second Mesa of the Hopi reservation upon which is carved their past, present and future history; it is known as Hopi Prophecy Rock. One petroglyph pictures the 'people' emerging from the underground at the beginning of the Fourth Creation and continues by showing how the clans migrated in the four directions, destined to return to the centre. The Great Spirit, called *Massaw*, called this centre *Oraibi* and said when the people who had scattered finally reach the real spiritual centre, the Fourth Creation will end. The petroglyphs show, symbolically, that if people remain 'two-hearted,' that is, thinking solely with their heads and not their hearts, it will lead to self-destruction. Patricia Mercier, who visited Oraibi, goes on to explain that the prophecies included three 'world shakings,' two of which are believed to have happened already. The second 'shaking' is believed to have been WW II in which "man used the reverse* of the Hopi migration symbol (the swastika)." The symbol for the third is called 'a gourd of ashes,' and will come after trees die, 'man builds a house in the sky,' cool places become hot and hot places become cool, lands sink into the ocean and lands rise out of the sea, the sea turns black and many living things die (Does anyone remember the BP oil spill of summer, 2010?), many spiritual seekers coming to learn our (tribal and spiritual) ways, and, the final sign, there will be an appearance of the *Blue Star Kachina*.[54] Ms. Mercier goes on to quote a Native American teaching, "Do not put one foot in the red man's canoe and one foot in the white man's boat since the ancestors foretold that a great wind would arise and tear the canoe and boat away from each other." She took this to mean that we (non-indigenous) really cannot understand their prophecies as it is a lifetime journey of study and they believe they have a sacred duty to preserve their spirituality. "Those who stay and live in the places of the Hopi shall be safe." The final stage of the prophecy, called 'The Day of Purification,' described as the 'hatching of the mystery egg,' and that of the return of the Great White Brother, *Pahana*, who has one of the four original Hopi Clan Rocks, are the most mysterious. The symbols on these rocks represent forces in a magnetic field, "and where they cross . . . is a vortex – like a stargate – into the next dimension."[55] (That word 'dimension' just keeps popping up, doesn't it?) Almost all of the prophecies point out: We still have a choice, the choice to respect and love all people as well as our Mother Earth. Graham Hancock tells the words an elderly Hopi elder of the Spider clan that he interviewed who said that the 'signs' are already here and "that this earth is the work of an intelligent being, a spirit – a creative and intelligent spirit that has designed everything to be the way it is . . ." "Nothing is here just by chance, nothing happens by accident – whether good or bad – and that there is a reason for everything that takes place . . ."[56]

The Hopi, which Ron Redfern said is shortened from *Hopituh*, "the peaceful ones," established the village at Old Oraibi around 1200 AD and have occupied it continuously since. He reports the Hopi believe the *Kachinas* go to the San Francisco Peaks between the summer and winter solstice and then to their own spirit world.[57] Ceremonies are performed to dispel the forces of disharmony in our lives and in the world. Nan Moss relates that a retired director of Indian Affairs observed that the Hopi say they do not make it rain with their ceremonies, rather they drive away the disharmony (called *akina*) that prevents the rain from falling.[58] Or, as Greg Braden says he learned from the indigenous people: "We do not pray for rain, we *pray rain*," in other words, they strongly visualize rain. Other examples of this type of summoning between worlds seem common for spiritual teachers such as Thich Nhat Hanh and the Indian Holy Mother, known as Ammachi, who frequently bring rain to an area they are visiting.[59]

The Desana of the northwest Amazon believe that the "cosmic and the human brains pulsate in synchrony with the rhythm of the human heartbeat, linking Man inextricably to the Cosmos," as discovered by ethnographer Gerardo Reichel-Dolmatoff when he studied this ayahuasca-using culture which believes the cosmos is centered on the middle star of Orion's belt.[60] The Maori of New Zealand say in 2012 the "veil will dissolve," meaning the boundary between us and where the ancestors have gone will disappear. Prediction, in the indigenous science, is not about knowing the future (linear flow of time) as much as a celebration of return and renewal, as the Mayans placing emphasis on the 'wheels of time.' In a society that views time as a circle, such as the Hopi, their concern would be "with that edge of manifesting where things pass from one world into another."[61] Here we have again the idea of *life constantly coming out of an unseen dimension, or matrix.*

As F. David Peat also wrote, "Instead of speaking of people having access to 'alternative' or 'nonnormal' realities, it is probably more accurate to simply say that Indigenous people live their lives in a wider reality."[62] 'Dreamtime' practiced by the Australian aboriginal people and Native Americans speaking to rocks are examples of this. The Indigenous people ". . . in South Africa in response to a crop circle appearing in the fields, would rush to erect a fence of poles around the circle . . . (and) would dance and perform sacred rituals honoring the star gods and the Earth Mother. ". . . their appearance would be cause for celebrations which lasted several days . . . accompanied by prayers to the gods to watch over them and talk to them through the sacred sites."[63] Silva also wrote, "The Maya believed that on the winter solstice (of 2012), an alignment between our solar system and the Milky Way will trigger a time of spiritual acceleration." ". . . the Aztec calendar shows we are at the end of a 13,000-year cycle and that significant changes in the Earth and in human consciousness are predestined like clockwork." ". . . the Hopi . . . prophecies state that in previous transitions a disconnect existed between humanity and the Earth, along with a polarization of heart and mind."[64]

"The prophecies say that we are entering an era governed by the element of *ether,* often described as the synthesis of - earth, fire, air and water – (it) is celestial, lacking material substance and existing beyond the confines of time and space."[65] Dr. Page goes on to explain that in ether there is a fusion of polarities that eliminates the need for separation into darkness and light, negative and positive, or good and bad. "In the new world, the unification offered by the element of ether requires the acknowledgement of unity through diversity where *every aspect* of the creative source is respected and honored without judgment, bias, or isolation."[66] This sounds like Einstein and the Holy Mother Ammachi rolled into one!

The Incas

At last we come to the Incas and the indigenous people of Peru and the man who has studied and worked among them most of his adult life: Dr. Alberto Villoldo, psychologist and medical anthropologist. His adventures in the Amazon and on the *Altiplano* and his training with Q'ero shamans, read like fiction, though all are true experiences. I treasure all his books. Cuzco, whose original name, *Quispicanchis,* meaning 'navel of the world,'[67] explains Patricia Mercier, was the seat of the wealthy kingdom the Incas spread over most of the west side of South America, and was one of the last great battles of Pizarro's Spanish conquest, which obliterated an entire advanced culture. Aided by an epidemic of smallpox and the common cold virus, Pizarro pushed on through *Sacsayhuaman*; the Inca leader, Manco, fled high into the Andes Mountains. The Incas, like the Mayans, based their activities, religion, their life, around the sun, moon, Venus and the Pleiades as well as the Southern Cross constellation. The Quechua word for star is/was *chasca*. It was reported by the Spanish invaders that the Incas had 328 *huacas,* or sacred places, openings in the *Pachamama*, natural sites such as rocks or springs or temples, where ceremony could be performed.[68] (Anthony Aveni)

The Incas also have a prophecy: Patricia Mercier met a *Chasqui*, or spiritual messenger, a descendant of the Quechua, who predicted that the end of the Inca calendar in 2013 will be marked by the appearance of a large magnetic asteroid. The *Chachapoya* or Cloud People of the Sacred Valley in the Peruvian high mountains, by legend, reportedly hold the Golden Sun Disc of Koricancha (sometimes spelled, Coricancha) which holds the mystery of world ages that was passed to them by the Atlanteans.[69] Large rocky outcrops were held in reverence by the Incas and their ancestors; some in the area of Cuzco have carvings on them. Mercier received a message from the crystal skull in her keeping while in Cuzco which said, "From 1992 to 2012 will come the merging of our three realms, Upper, Earth, and Underworld (*hanaq pacha, kay pacha* and *ukhu pacha*). This time period is called the *Taripay Pacha*, which, translated, means 'the age when we meet ourselves again." "We are the Q'ero, the last of the Inca, and we ask you to look with the eyes of your soul and to engage with the essential. Regaining your luminous nature is a possibility today for all who dare to take the leap."[70] Ms. Mercier recalled that Hunbatz Men (the Mayan leader of whom we have spoken) told her that the *Q'ero* believe doorways between worlds are opening again, "like holes in time that we can step through and beyond, . . . places where we can explore our human capabilities in luminous bodies." [71]

The modern Incas believe that beginning at the festival of *Qoyllur Rit'i* which takes place near the glacier on Mount Ausangate (20,594 feet elevation) in the near future, six males (*Inka mallku*) and six females (nust'a) of the fifth level (this level of consciousness allows miraculous healing) will meet. The 12 Shining Ones will return to the Temple of *Wiraqocha* in Cuzco where two will attain the sixth level, will have a visible aura,[72] and open the golden age in which it will be possible for the seventh and highest level of consciousness to arise in humanity. This was revealed to Barbara Jenkins, a fourth-level initiate of shaman, Juan Nunez del Prado, who believes the fifth level of consciousness will have to become established before the Luminous Ones on the sixth level begin to increase. The end of the whole transitional process would end on August 1, 2012, the first day of the Taripay Pacha and the Q'ero New Year's Day.[73]

As Villoldo's teacher/shaman, a Quechua Indian, said, (certain) ". . . symbols are universal. You find them in every culture in the world . . . . The shaman knows that there is a sea of consciousness that is universal . . . an awareness and a world we all share. . . And the shaman is the master of this other world. He lives with one foot in this world, and one foot in the world of spirit."[74] This wise master teacher (a *haitun laika*), don Jicaram, taught him the prayers to the Four Winds (the cardinal directions) and reverence for the Earth and Sun, saying, "And like a newborn infant who knows where to find his mother's breast, the Children of the Sun were born with the knowledge of their origins. That is why this is the first story ever told. It is the first things known. We tell this story around a fire, because whenever we build a fire, we celebrate life by remembering the time when the Sun shed a tear of joy that landed on the pregnant belly of the Earth . . . and the Sun and the Earth would go on making life, for they are very passionate. . . the Children of the Sun recognized the gold on the ground . . . and practiced the alchemy using the fire of the Sun and the living soil of the Earth to create . . . a living gold – corn. Others, who called themselves children of a god, coveted the gold of the Earth and killed the Children of the Sun, the Incas, but they (the Incas) left their bodies behind and went to Vilcabamba. To this day, the journey East, to realize the visions and skills acquired along the Journey of the Four Winds, is literally the return home – to the place whence we came – the source of life and the creative principle – Vilcabamba, "not a place (you) can walk to . . ."[75] He taught Villoldo how to make *ayni* (ceremony, prayers, praise, gifts) to the ancestors, the storytellers, the caretakers of the Earth, the Four Winds, and to *Pachamama,* the Mother Earth. The principle of sacred reciprocity (mutual exchange) is the basis of Andean shamanism. The ancestors of the Incas "worshiped the Sun and the Earth . . . they knew that there would be no life on Earth if it were not for the Sun, that life is the direct result of their union – and their parents were one – *Illa Tici Viracocha* – neither male nor female, energy in its purest form." ". . . our world is always a true reflection of our intent and our love and our actions." "The condition of our world depends upon the condition of our consciousness, of our souls."[76] The creation story Villoldo tells from don Jicaram at the beginning of *Island of the Sun* will bring tears to your eyes. He relates how he was taken into the 'fold of the Laika,' a Wisdomkeeper tradition that predates the Incan civilization. Beginning with his association with Professor Antonio Morales of the Universidad Nacional de San Antonio Abad del Cuzco, Villoldo came to know him as the master medicine man, Don Jicaram.

Dr. Villoldo spent time at Machu Picchu with don Jicaram and other shamans and felt it was one of the places that had always been sacred – "places that lend themselves to meditation and transformation . . . or simply sings to our unconscious."[77] The Gateway of the Sun, *Intipunku,* is the entrance to Machu Picchu, the Sacred City of Light, which lies 2,000 feet above the Urubamba River and between twin peaks of the same mountain, Huayna Picchu (Grandmother Peak) and Machu Picchu (Peak).

Some say the city was built by the Incas, others assert that it existed long before them. In a courtyard is the *Inti Huatana*, 'the hitching post of the Sun,' where the Sun was tied by the priests on the occasion of the winter solstice to keep it from going further north and be lost.[78][79] In the ruins the Pachamama stone has been the site of many initiations and 'journeys' especially in retrieving the feminine side of our nature. Shaman, don Eduardo said that a shaman is not "able to 'see' with his or her inner vision until his/her feminine side was awakened." As Villoldo said, "Shamanism and modern science seem to approach knowledge from opposite ends of the spectrum, yet both provide us with a coherent image of reality.[80] Freddy Silva (the crop circle researcher) expressed the same knowledge, "Scientist and shaman, it appears, are aboard separate buses en route to being reunited at the bus station." ". . . the most wondrous elements in the Universe – love and inspiration – continuously fail to be measured, scientifically proved or quantitatively analyzed, so according to science these states shouldn't exist."[81]

Legends speak of Machu Picchu as "the doorway between worlds, where the veil that separates us from our dreams and the stars becomes thinner, diaphanous, easily drawn aside."[82] In a film, Villoldo says, "There will be a tear in the fabric of time itself, a window into the future through which a new human will emerge . . . Homo luminous.[83] He also believes the Star Rites held each year on Mt. Ausangate may bring forth the emergence of the 'luminous ones' – 'the ones who walk in peace.' Ancient stories also say that the Andes were split apart "when the sky made war on the earth."[84] This is believed to have happened at the time of the great catastrophe. The myths surrounding the Viracocha are numerous. This name is used as the one Great Creator God, as well as the tall, bearded wise man that came after the 'great deluge' and taught the ancestors engineering, architecture, and skills in medicine, farming, road building and metallurgy, as well as peace – a golden age. The walls, buildings and gates of granite boulders weighing more than 100 tons, in the ancient parts of Cuzco, Sacsayhuaman, and Tiahuanacu attest to an expertise (that some call technically advanced or even miraculous) not known to exist at that time. Graham Hancock reports that the joints between the stones are cut so precise a sheet of paper cannot slip between them.[85]

Viracocha and his attendants (called *hayhuaypanti* or 'the shining ones') are also credited with one of the most enigmatic discoveries of South America – the Nazca Lines. The Nazca Plateau of southern Peru is one of the driest places on earth, so not much has changed since these enormous designs were placed on the gypsum soil, some say, thousands of years ago. The designs of birds (18 different), animals, human forms, parallel lines, geometric shapes and the famous spider figure (said to track the Orion constellation) are spread over a 200 square mile area. Because of their sheer size, some 100 to over 600 feet across, they are best seen from the air, appearing to have been executed by a giant, Hancock reports. This largest piece of graphic art on earth was made by scraping away thousands of tons of black volcanic pebbles to expose the desert's base of yellow sand and clay (too soft, according to one researcher, to permit the landing of alien spacecraft which has been postulated!)[86] How does this mystery correspond to our examination of dimensions and signs? Better question: How did the 'person' or group with the giant plan relay this to the hundreds or thousands of people required on the ground to execute the moving of the mass of topography? Did the design come from the same 'dimension' that healers, such as John of God, pull from?

Healers

Oprah Winfrey devoted an entire show to João Teixeira de Faria, known as John of God, or Medium João, of Abadiânia, Brazil. Thousands have come to him; hundreds arrive here daily from all over the world to be healed, but João says, "I have never healed anybody. It is God who heals." His services are free of charge. For over 52 years, since he was 16, about thirty Spirit entities (disincarnate doctors) have worked through him, mostly without incision (those with incision are never sterilized, always without anesthesia, without pain), to cure cancer, tumors, diabetes, bone fractures, blindness, hearing loss and the entire gamut of physical and mental/spiritual afflictions. Physicians, scientists and non-believers have come away from Casa de Dom Inácio de Loyola with a changed view of reality. "Brazil has deep roots in the traditions of shamanism and spiritism, both of which feature the notion that individuals can – and do – cross the boundaries between this worldly existence and the afterlife." Susan Casey travelled to Brazil for Oprah to bring back this story.[87] Ledwith and Heinemann have photos of 'orbs' at a healing by John of God as well as at events of other spiritual healers. They believe that these helping Spirits are immensely intelligent, acquire information at infinite speed, but do not act until a readiness is indicated within the thoughts of the living. As they say, Spirit emanations can go within the body and selectively correct with billions of energetic interventions at the cellular level within fractions of seconds; the healing ability is limitless, but happens only upon request.[88]

Many spiritual teachers have maintained that consciousness of infinite range in this 'unseen dimension' is "available to tap into *at any point in time* by any being with a capability and will to do so" as taught at the Oneness University in southern India and by the *Abraham-Hicks* materials,[89] both of which we will look at in depth in the next chapter. Time does not seem to exist in this 'dimension' since telepathic communications were shown to happen weeks before in experiments José Argüelles conducted with scientists at Novosibirsk University in Siberia. Information about Mayan temples, symbols, hieroglyphics and artifacts were sketched in April, 2005, though the formal transmitting of this information did not begin until late May, 2005. The scientists concluded that psychic phenomena defy conventional cause-and-effect notions of time, as has also been shown by animals that sense earthquakes and tsunamis hours or days in advance. Some have explained this as sensitivity to geomagnetic disturbances or in the case of psychics, intuition (but from where does intuition arise?)[90] The hundreds, perhaps thousands, of books about angels should give us a hint that there is more than what we can see, taste, touch, hear and feel.

A personal experience here will show why I cannot discount this area of the 'unseen worlds.' When my previous husband was in Hospice care in the final stages of lung cancer, I knelt by his bedside one night just to be near him in his semi-comatose, transitioning state. I became aware, 'saw' (with my mind, not my physical eyes) that there was a 'group' of 'spirits,' what I would have called 'cherubs' (angels) over his head, sending him so much love, encouragement and, yes, joy, that it flowed over onto me, though they seemed to pay no attention to my presence. He died twelve hours later, but that 'vision,' if you will, helped me tremendously through the following months of grief. I received such peace from knowing that he had this wonderful 'cadre of loving friends' helping him 'cross the veil.' I am not psychic, and have had only one other 'experience' of this type, yet I feel I can speak about this part of that other 'dimension' beyond the 'veil' with some confidence. Solara talks about angels in her book, *The Star-Borne,* "Angels are known to Moslems as *Barakas*, to Chinese as *Schiens*, to Hindus as *Devas* and to (some) Native Americans as the *Bird Tribes*. Throughout recorded time, Angels have appeared to humanity as Messengers from God and instruments of Divine Intervention."[91] Doreen Virtue, Ph.D. is widely known for her many books about angels.

Crystal Skulls

Patricia Mercier relates in her book about the crystal skulls, *The Maya End Times*, ". . . this light (radiation) prepares Earth for a 'birth' which will occur through the Dark Rift of the Milky Way . . . because when (it) is in this precessional alignment . . . a much more spiritually evolved cosmic energy radiation will be received over the whole planet . . ." "Children will intuitively understand how to utilize this energy and quickly be able to reach a superconscious state, transiting at will into the fifth and sixth dimension."[92] You may have heard of Indigo Children and the increase in autism.

Hunbatz Men, Itzá Maya spiritual leader, whom we met earlier in our 'journey,' wrote in the Preface to *The Maya End Times*, "The crystal skulls are very important . . . and the need . . . to understand (them) has become urgent." He says the ancient Mayas had access to a more advanced knowledge by their worship of the crystal skulls. "The most perfect crystal skull . . . was found in the Maya temples of Lubaantún, Belize, by Anna Mitchell-Hedges." "When all the crystal skulls around the world work together, it will be the right time for the beginning of a new age . . . an era of wisdom."[93] Ms. Mercier has travelled all over the globe and wrote that the skulls are fundamental to the prophecies of the Maya. "Mystics and visionaries who have held the crystal skulls have accessed information that can only be explained as coming from other dimensions." Didn't I warn you we were going on a very mysterious 'journey'?

Ms. Mercier was given one of the crystal skulls by an elderly indigenous woman as they meditated together in a *kiva* at Chaco Canyon (Anasazi ruins in New Mexico). She went on to meet with several other people who were guardians, or caretakers, of other crystal skulls, and was asked to join in a ritual at Teotihuacán with her skull. The indigenous people who she encountered told her, "Traditional inhabitants of Turtle Island (the Americas) believe that crystal skulls were programmed by the ancients under the guidance of extraterrestrials (from the Pleiades)*. They used light, much in the way scientists now program the very small, clear and perfect silicone quartz chips used in all our modern technology." "I could just imagine the quantity of information that could be held within a 'chip' the size of a skull!" she wrote. Others asserted the skulls are thousands of years old and are 'record keepers.' One woman shaman told her, " . . . coming into flesh, we forgot how to see the luminous fibres connecting us from the part of our spiritual body that we call the assemblage point, to the energetic pathways and the WorldWideWeb of Life and Light."[94]

The Mitchell-Hedges crystal skull (called a 'singing skull by the Mayans) was examined and tested by Hewlett-Packard in California, who confirmed it was made from a very pure type of natural rock crystal, silicon dioxide, only slightly softer than diamond, identical to that used for chips in computers. The scientists there asserted that making such a skull would be near impossible, even with modern cutting techniques, and the skull had no evidence (tool marks) of any modern instruments used on it, even under great magnification. They were at a loss to explain its origins, saying it would have taken over 300 man years of work to complete! The skull also had unusual optical properties: light focused from below came out the eye sockets. Two other skulls, one of amethyst, were tested by HP and also found to be inexplicably cut against the axis of the crystal. Since quartz is used in time pieces, it follows that the "skulls can hold or process information in the form of electronic pulse, just as silicone chips in computers and telecommunication satellites do," Mercier reports.[95]

One Mayan legend relates that there are 52 ancient skulls that came from the Pleiades star system and hold secret keys for humanity's greatest leap of consciousness. Another says the skulls were brought here from Atlantis 13,000 years ago. Many of the skulls in private collections, such as those found in a cave in the Himalayas, are being actively worked with in meditation and ceremony. Some of the skulls have elongated craniums similar to those seen in ancient Egyptian tomb paintings. Some modern Mayan shamans and elders, upon holding one of the skulls, have enhanced abilities to read the 'Threads of Time,' and say they (the skulls) are teachers.

Several caretakers have taken the crystal skulls to certain power points on the planet to heal the Earth's grid and restructure damaged energies. Mercier wrote that the skull in her care had shown scenes of the Earth suffering saying, "Everything you do to one you do to all." "You aren't listening to the rain, the wind, the oceans, volcanoes, earthquakes and tsunamis."[96] Many of the indigenous prophecies decry the 'rape' of Mother Earth as incestuous by her children, and say that war has grown out of greed. I am reminded of the similarity of the words 'inequity' and 'iniquity.' As Mercier says, "ancient, spiritually based communities such as the Hopi must especially be preserved, without being forced to abandon their ways of living and the natural resources they have vowed to protect." A Cherokee elder, 'Metis' Thomas Thunder Eagle relayed this sacred message from Elder Heyoehkah of their council: ". . . this is why we have left these receptacles (crystal skulls) for you to find . . . long, long ago. . ." "There would be those who would be needed to call upon their reincarnation memories to heal, to counsel, and to love a world gone mad, a world without knowledge, a world without hope, where the fires of destruction would reign. Within this (crystal skull) and others we have left you lies that which you will need. It was determined that, through these receptacles, the minds of oneness would be activated and would present themselves when your earth was in need.[97] In Appendix 5 of *The Maya End Times* Ms. Mercier lists the most well-known crystal skulls 'discovered' in the last four or five centuries in the Americas.

Chris Morton and Ceri Louise Thomas also undertook far-reaching research of the crystal skulls and wrote of another Cherokee legend which says, ". . . there are twelve planets in the cosmos inhabited by human beings and that there is one skull for each, plus a thirteenth skull vital to each of these worlds."[98] They visited Anna Mitchell-Hedges, discoverer and caretaker of the crystal skull from Lubaantun, and were very impressed with its perfection. Ms. Mitchell-Hedges believed her good health and long life were because of the skull; she felt it had always protected her. This skull weighs almost 12 pounds and if the sun's rays are directed through it in a certain way can start a fire on paper, like a mirror. Anna's father, Frederick, an explorer and archaeologist, (whom Anna was with when she was 17 and made the discovery) believed that Lubaantun was *pre-Maya*, because the building techniques were different from other Mayan sites, more like the

Incan city of Machu Picchu. The Mayan priests told the Mitchell-Hedges that the skull was 100,000 years old, that it was a 'healing skull,' a gift from the Mayans to mankind, and was very important to this time.[99]

They visited with several other people who were 'caretakers' in North and South America finding that some of the skulls triggered channeling of information in some people. One said, ". . . it was made many, many thousands of years ago by beings of a higher intelligence . . . before those you call 'the Maya.' It was molded into its present form by thought. The thoughts and knowledge are crystallized into this receptacle. "Crystal is a living substance." "The Earth life of this receptacle is 17,000 years." . . . "There will be other receptacles found, for there are many. . . one where we left markings upon the Earth [some have suggested Nazca Lines] and high in the mountains. There will be one of blue in the region you call 'South America.' There will be another found when the lost civilization that you call 'Atlantis' rises to you." ". . . at this time . . . it would be too dangerous for man to have this information . . ." (It is) too early in his evolution. . ." "The receptacle has been given this form to encourage the mind of oneness and to reduce you desire for separation. . . (this) causes annihilation and death. . ." "There is much violence occurring – violence against men, violence against nature . . . violence against the Earth." "In that which you call deserts you will find much knowledge. . ." "We can only permit that which will not cause too much havoc for your minds. . ." "From other dimensions we came into this dimension." ". . . you still cannot grasp the idea that we are of the other dimensions, beyond your rudimentary space-time relativity." "The essence of time is an illusion. . ." "Thought is without time." "We have come now to warn you, because the separation . . . (and) destruction is beginning. There will be more destruction brought about by atmosphere intervention . . ." "Your scientists and governments play . . . with light and sound and that which you call 'particles' and 'radiation' and they will rain havoc." ". . . this is why we left these receptacles, when we realized that so many had forgotten their original purpose of incarnation into this physical dimension." ". . . although what is given cannot now be changed, its effects can be much diffused."[100] And that is what we will talk about in the chapter on Consciousness.

Morton and Thomas in their search for the crystal skull that was rumored to be with the Navajos, talked with a Navajo spiritual leader who explained that the skulls were very sacred to their people, only the 'holy ones' were allowed to know where they are, and that the shaman who looked after the skull in the mountains would be the only one to know where it was. He later told them, after speaking to the spirits and other tribal elders, that they were called 'the skull people' when the Spanish first arrived in the area. Echoing the words of Chief Seattle, he said, "The land does not belong to the people. The people belong to the land . . . the land of the skull." He showed them a map and noted the 'sacred crystal mountain,' but said no one should look for it. "Such things are best left alone. . ." "Take comfort in knowing it . . . is fulfilling its purpose and is being looked after carefully."[101] The Navajo believe that turquoise facilitates communication with the spirits. The authors were invited to walk with the group on their yearly pilgrimage to the place where their ancestors emerged after the flood had destroyed the previous World; the petroglyphs there depicted their creation story. The elder disclosed that the skulls, made by the 'Holy Ones' were a template for the whole of humanity, a blueprint; that there is another level on which things exist that is invisible to the naked eye. "But it is on this level that all things were first made. This other dimension contains the dimensions and definitions for all . . . species and even mankind." (This sounds very much like what Professor Tiller called a template in his research!) "The skulls are like a map of our human potential. They are there to help humanity to steer the right course, to help the mountains remain intact . . . and even to ensure that the continents stay in their current place. . ." "Everything is just a vibration, constantly moving and changing . . . our very consciousness is vibration. Crystal is vibration . . . a particular type of vibration that sings to our spirit . . . (and) we can hear it with our hearts. This is why the skulls are sometimes called 'the singing skulls.' Their song is the joy of creation, the wonder of existence." "They remind us that we are part of the miracle of creation and this sound can help us steer the right course through space and time."[102]

The Native American author Jamie Sams told Thomas and Morton that the skulls were multi-dimensional and could teach us about transformation. At the Santuario de Chimayo, a Catholic church built on a sacred indigenous site, she continued, "Quartz crystal . . . can quite literally connect us with the workings of the spirit world. There is actually only a very thin membrane, like tracing paper, separating this physical world of ours from the world of the spirits. Indigenous people all over the world have long believed that quartz crystal can provide an access point into this spiritual dimension...." "The skulls represent the Native American faith in the intangible force that all religions call God, the Creator, the Maker, and we call the Great Mystery or Great Spirit." In the side room of the church is an opening revealing the 'healing sands': tiny particles of thousands of broken down pieces of quartz crystal. "Just as the crystal skulls were believed to have healing powers, so too were these fragments of quartz."[103]

These authors were invited to attend an important gathering of Indigenous elders from North and South America at the Great Pyramid of the Jaguar in Tikal. Amid copal incense, the high priest of the Council of Quiché Mayan Elders, Don Alejandro Cirilo Oxlaj Peres (his Spanish name), or Job Cizin (his Mayan name for his birth day sign: 'Life/Death') said, in short, "We are not the first messengers; the first messengers are the crystal skulls..." "The prophets taught the great science that is called 'the Mayan Light' . . . (which) says 'Let the dawn come . . . for the love must not only be between humans, but all living things. We are the children of the sun..." "We are all one like the colors of the rainbow . . . the Warriors of the Rainbow are now being born . . . and it is time to prepare for the Thirteen Heavens. Now is the time that the prophecies will be fulfilled."[104] Don Alejandro reiterated that there are 52 skulls and 13 were left with the Mayans and that was one reason these numbers were sacred and played a part in the Mayan calendar.

Apocalyptic Predictions

The word 'apocalypse,' has the negative connotation of doom, but actually comes from the Greek word *apokalypsis* meaning 'to disclose or reveal.' In *The Isaiah Effect* Gregg Braden writes that *The Essene Book of Revelations* found in the Dead Sea manuscripts when translated from the native Aramaic language is very similar to the version of Revelations in the modern Bible. John's fragmented vision of chaos, death, terror and destruction causes him to ask, "Is there no hope?" To which the voice replied, "There is always hope, O thou for whom heaven and earth were created," and further asserts, "And I saw a new heaven and a new earth . . . I heard a voice saying there shall be no more death, neither sorrow, nor crying, for the former things are passed away." "Nation shall not lift up sword against nation, neither shall they learn war anymore. . ." The Isaiah Scroll, the only manuscript discovered completely intact in 1946, now protected in a museum in Jerusalem, reveals a common theme writes Braden: "descriptions of catastrophe are immediately followed by a vision of life, joy and possibility."[105]

Another prophetic finding called the Bible Code was discovered by an Israeli mathematician, Dr. Eliyahu Rips, in the first five books of the Hebrew Bible, the Torah, written over 3000 years ago. Sophisticated search programs were used, after spaces and punctuation were removed, to examine the remaining letters and Michael Drosnin described these prophecies which are precise and appear to be mostly accurate, in his book, *The Bible Code*. Many events, from the John Kennedy assassination to the Gulf War, names and dates, have all proven correct. High speed computers have shown that none of the other books examined have these encryptions.

The prophecies relate to our future and "infer that all possibilities of all futures are already in place" – like a hologram.[106]

Braden suggests that the key to applying this ancient code may be in viewing it through the eyes of a quantum physicist, who knows that it is impossible to know the 'when' of something and the 'where' of the same thing, at the same time. If you measure the *where* you lose information about how fast it is moving. If you measure how fast it is moving, you cannot know with certainty *where* that something is; this is known as the Heisenberg uncertainty principle.[107] The possibilities of world war and catastrophic earthquakes are predicted there, as well as that of a comet near-miss in 2006 which did not happen. For the year 2012, there is an unsettling sequence of words: 'Earth annihilated,' and a second passage says, "It will be crumbled, driven out, I will tear it to pieces, 5772" (the Hebrew year for 2012).[108] Similar paradoxes show up throughout, followed by the same four words again and again: "Will you change it?" This suggests that **we** play a significant role in the outcome of events, even those set as possibilities. Each prophecy stops short of detailing how this Great Cycle will end and the next begin, acknowledging the "power of mass choice expressed as the science of mass prayer." Then Braden quotes from *The Essene Gospel of Peace*, "And one day the eyes of your spirit shall open, and you shall know all things." [109]

The word "Armageddon" appears only once in the entire Christian Bible writes Sylvia Browne and yet "has become a dreaded synonym for the final battle between good and evil, Christ and the Antichrist, God and Satan . . . that will end the world"[110] I know of some fundamentalist, evangelical Christians who believe that Armageddon occurred a few decades after Jesus died. They see Revelations as a fantasy given to John, the apostle, by God. I think he may have suffered from a lack of entertainment in his exile on the Island of Patmos! (Not to be unfeeling or sacrilegious!)

From the television production recently aired on CNBC (April, 2011) about Apocalypse 2012, I learned there are quite a number of people who are making serious preparations to survive a worldwide catastrophe in December of 2012. They related fears that the wandering 'planet' Niburu with a 36,000-year cycle would return, or that the governments of the world know about the dangers and are keeping it from the masses. One family turned their pool into an underground storage/hideaway. Another man plans to jump off Bell Rock in Sedona, AZ; he says a portal in the sky will open and catch him in his 'leap of faith.' I was amazed at the bunkers (one in Kansas in a missile silo that is 14 stories deep), buried shipping containers for homes, stockpiles, weapons and remote underground dwellings being amassed in several countries. A NASA scientist, Morrison, calls it cosmophobia and thinks believing all this will create mass hysteria and bring on a self-fulfilling prophecy. When you think about how consciousness works this could be true; it seems better to believe that we are capable of pooling our positive thoughts and actions to create a future that would have little need for 'bunkers.'

Sylvia Browne quotes from the Inca Q'ero (Incan) shamans:

> "Follow your own footsteps
> Learn from the rivers,
> The trees and the rocks.
> Honor the Christ,
> the Buddha,
> your brothers and sisters.
> Honor the Earth Mother and the Great Spirit.
> Honor yourself and all of creation.
> Look with the eyes of your soul
> and engage the essential."[111]

Dr. Carl Johan Calleman quotes the Qur'an, "Again, what will make you realize what the day of Judgment is? The day on which no soul shall control anything for (another) soul and the control shall be entirely Allah's." (Cleaving Asunder Surah 82:17-19) Also the Hopi prophecy: "The spiritual beings will remain to create one world and one nation under one power, that of the Creator."[112]

It is not an oversight that the prophecies of the Christian Bible have not been examined here to any extent. To me, it seems that when that book was edited in 325 AD by the Nicene Council most of the Gnostic viewpoint was left out, and it has been mistranslated and misinterpreted to the point of almost losing Jesus' main teaching of Love. I grew up in the fundamentalist Protestant setting, attended a Christian college, but could not "fit into' that point of view after studying Eastern philosophy and religions, Native American spirituality, the power of the mind, and the Dead Sea Scrolls; even the poetry of Jelaluddin Rumi (Islamic Sufi, 1207-1273, almost contemporary with Francis of Assisi, (1182-1226) changed my views about our connection to Source and how we really are all the same at a deep level. How could one not be changed, after reading Coleman Barks translation of Rumi:

> There is a force within that gives you life –
> Seek That.
> In your body there lies a priceless jewel –
> Seek That.
> Oh, wandering Sufi,
> If you are in search of the greatest treasure,
> Don't look outside,
> Look within, and seek That.[113]

Modern Catastrophes and Predictions

Upon mentioning the Mayan calendar and 2012 at a medical appointment recently, I received the 'typical' response, "Oh, the end of the world." To which I replied, "No, my belief is: that depends on how much consciousness can evolve in the next few years." So, are there any more 'signs' that we need to look at before we journey on into 'Consciousness?' We have looked at science and what Einstein called 'quantum weirdness' now being approached by the theory of 'entanglement' in which a photon, split in half, (That's the trick I wanted to see!) each is shot in opposite directions, and each part reacts to choice or modification identically, instantaneously, as though still 'one,' though miles apart. Their connection was absolute. Gregg Braden calls this infinite web of connection *The Divine Matrix* and is the name of his book on this subject. He suggests we look at the time before The Big Bang knowing at that time everything we know as the physical world was connected in a very small space and is still connected through this field of energy that Max Planck (famous physicist) called the 'matrix.' He goes on to relate this all-encompassing connection to time and consciousness.[114] The Abraham-Hicks materials offer, "There is a current of Energy that surges through the walls of your Universe and you are already plugged into it. The way you utilize it towards your creative end is by contouring it through thought." (Abraham – G 4/6/91)[115]

I like referring to this great bond as The Web or The Tapestry (part of which we each weave with our lives) and believe it is what we can affect and shape with our thoughts and positive intentions. As I write today, the so aptly named WorldWideWeb of information tells of the 9.0 earthquake near Japan that has caused a huge and damaging tsunami there. This is reported to be the fifth largest earthquake the world has seen since 1900. We are all touched and we want to reach out to our fellow humans; we can feel their panic.

The prophecies all predicted some catastrophes, and in the past few years we have seen what seems to be an increase in these worldwide. One anomaly that has come to my attention, what the newspapers called the 'Aflockalypse,' is the sudden death of so many birds, great flocks, hundreds to thousands at a time all over the world, seemingly worse in the Northern hemisphere. Add to this the tens of thousands of different types of fish dying in masses from Thailand, China, New Zealand to the US, UK, Brazil and Canada. Not all this can be marked up to 'better reporting' worldwide, nor are pollution, parasites, virus, or magnetic field aberrations enough to explain the sheer size of this. In 2003 there was a large earthquake in Iran; in 2004, the 9.1 earthquake near Indonesia with the resulting tsunami is estimated to have killed over 300,000 people, and was followed a few months later by an 8.6 in the same area. The phenomenal hurricane season in the U.S. in 2005 now seems mild compared to the volcanoes, earthquakes, floods, tornadoes and snow storms that rocked the world in 2010. One victim of the January, 2010 earthquake that devastated Haiti said, "The world is coming to an end." We can certainly understand her feelings when we hear that around 300,000 people died in this disaster and the disease that followed. Now the earthquake/tsunami in Japan has resulted in over 20,000 people dead or missing and the Fukushima Nuclear Plant destruction threatens a large area. The Hopi 'Time of Purification' seems to portend all of this rather well. The U.S. saw in 2011 an especially destructive tornado season, causing over 500 deaths, well over 100 in Joplin, Missouri alone. Do we as a society need to visually see all these catastrophes to help us change our apathetic viewpoint?

Geoff Stray has pulled together much information on topics about 2012 in his book and on his website[116] and writes about a modern prophecy that is emerging from a different source: those coming from NDEs (near-death-experiences) and OOBEs (out-of-body-experiences).[117] Research from several different fields is showing a correlation between sunspot cycles, the geomagnetic field of earth and human biochemistry, specifically the pineal and pituitary glands. Neuroscientists have shown that a change of as little as 20 nanoteslas (that is a billionth of a tesla) can precipitate a seizure in people who suffer from epilepsy. During solar storms or flares our geomagnetic field can change by as much as 500 nanoteslas! Sharp changes in this field are known to affect the pineal gland almost like the effects of harmala alkaloids in various hallucinogens. Under the 'right' conditions, the pineal gland can produce beta-carbolines and pinolines that can then release 5-meo-DMT, which will, of course, alter one's state of consciousness or cause an out-of-body or a mystical experience. People entering this state either by accident or self-induced (by using Syrian rue, acacia, psychedelic mushrooms, or other plants) seem to tap a common information source and bring back similar messages about 2012: that time will be perceived 'differently.' Some say geomagnetic changes will open our chakras, including 'The Third Eye (near the pineal gland), while others believe a mass OBE could occur and we would all be changed, like those that have been transformed by their experience of this kind.

Lazaris (pronounced Lah-zär´is), channeled by Jach Pursel beginning in 1974, revealed "The energy of your holographic universe flows through the Vortex of Sirius." (This constellation just keeps coming up!) "This star, Sirius – the brightest star in your heavens – was known by the ancients and is known by the 'primitive' societies of your modern world. The Dogon tribe in northern Africa . . . and aboriginal tribes in Australia, South America, and Asia also have knowledge of this star . . . (and) of its 50+ year elliptical cycle and of the dwarf star Sirius B." "July 23rd is the day when Sirius rises just before the Sun . . . (and) begins a 55-day cycle when the Vortex is opened more fully than at other times." ". . . the energy is more available and potent. . ." (There is) "a celestial alignment that only happens every 90,000 years, (and) in 1994 Light flooded into . . . and saturated the Earth . . . (and it) penetrated to the core – everything is different now." "Through it flowed the Light and Love that can allow you to transcend what has been and to create what will be."[118] In *The Sirius Connection*, this entity (or entities) also describes meditations to strengthen the hypothalamus, pituitary and pineal glands to help open the 6th and 7th chakras and help work with the influx of electromagnetic energy. We examined in the chapter on Astronomy this increasing 'plasma' energy.

Lazaris goes on to explain, "Because the Vortex opened – because intention and attention are more intense – you can drop the past." "It will be different." ". . . each of your chakras has been changed. . . as though the settings have been recalibrated." ". . . electromagnetic energy has intensified marked by greater tectonic activity of earthquakes, by volcanic eruptions, windstorms, electric storms." ". . . such energy and its range of manifestations will continue to increase. . ." "This energy called electromagnetic is essential to the evolution of your brain . . . the evolution of the human species. In short, you need this . . . energy." And from the Preface Lazaris says, "We are here to remind that pain and fear are not the only methods of growth, that you can more elegantly grow through joy and love . . . (and) that you do create your own reality."[119]

Ken Kalb's NDE described in his book, *The Grand Catharsis,* led him to 'see' a near-death-experience for the entire planet. Dr. Callaway of Finland has, by means of blood plasma spinal tap readings, shown that NDEs and OBEs are based on a pinoline and DMT release by the pineal.[120] Geoff Stray believes this 2012 information is "bubbling up from some deep, communal level of mind. . ." (the mass unconscious?) He also relates that Mazatecs today using a sacred plant, *salvia divinorum* (*pipiltzintzintli* in Nahuatl) have received more information about 2012.[121] Did Edgar Cayce tap into this collective unconscious (astral plane) when he said the Sphinx and the Great Pyramid were built in 10,490 BCE and would lead us to information we would need for 2012? I have not been able to find any specific predictions (or incidents of remote viewing) for after 2012, (with the exception of Sylvia Brown's *End of Days*[122] that predicts earth disasters until 2030), just generalities about 'a golden age' for those who survive and are able to enter a higher state of consciousness. Sylvia Brown includes in this book prophecies from the Brule Sioux, Cherokee, Catholic (*Parousia* – the second physical manifestation of Christ), Islam, Hinduism, Mormons and Zoroastrians. She also reiterates that the Mayans never intended to imply a cataclysmic end on 12/21/12, but that it was "humankind's choice as to whether that transition would involve violently dramatic changes or simply evolve with graceful, peaceful tranquility."[123]

Mike Dooley (featured on "The Secret") reminds us in his DVD movie "2012: A Wrinkle in Time"[124] that predictions and prophecies from the past (or any time) were rooted in that culture, religion and time and the prophets undoubtedly 'filtered' what they saw in their visions through their own 'window of reality.' Because of their particular mindset, they may not have realized that "nothing is predetermined," as Dooley says. If it were, teachings such as 'believe and you shall receive,' and the Law of Attraction would not be valid. If anything were predetermined, that would take away 'free will,' and most of us truly believe we have free will. So, in a sense, the prophet 'taints' the prophecy with his own beliefs, such as John, in Revelations in the Bible, must have believed that God was vengeful. Dooley further emphasizes that prophecies are probabilities or possibilities based on what the prophets 'saw' in a future scene and they perceived this through their own lens of experience and perspective. The Mayans seemed to realize or 'see' that the consciousness of humans would evolve, even metamorphose, that with this growth would come some chaos, but not more than humanity could cope with, since they predicted the Fifth Sun would be one of peace and harmony.

CONSCIOUSNESS

The Burgeoning Consciousness Movement

From all that we have examined, it becomes apparent that 2013 could be the beginning of a new way of Being. This change in consciousness may be 'assisted,' or possibly challenged, by physical changes to our surroundings, such as electromagnetic waves, plasma increases, and alterations in the magnetic pull of the sun, planets, and center of the galaxy. We are seeing, from recent events, some of the balancing processes of Mother Earth, and because of these changes, we may begin to perceive Life differently. Perhaps, we can learn from the Mayans; they understood the rhythmic and cyclic nature of the universe and life, and our connection to the eternal and infinite source-consciousness from which the manifest world springs. The *Tzutujil* Maya believe Spirit (*k'ex*, essence) and matter (*jal*, form) unfold together with spirit having priority.[1] Maintaining that important balance between ego and Self (essence), may be the challenge for consciousness today, just as it was for the Mayan shaman as she/he tried to sustain the conduit between earth and sky, between this world of form and the world of Spirit. Though we have examined some scary predictions and have already seen some catastrophic 're-balancing' (such as tsunamis and earthquakes) of the Earth, I believe, like John Major Jenkins, "we should not pour our energy into envisioning ('worst-case scenarios') and projecting them into manifestation with our fears."[2]

How do we not become fearful when we see scenes of massive destruction by a tsunami or a war on our TV? Information about consciousness has been pouring into our knowledge pool since the 1960s and gradually more and more has been picked up by the mushrooming mainstream media.

A lot of research has been done in the past forty years on subjects from psychology to neurological studies of brain waves. The Human Potential Movement, mind-over-matter studies, and the transformational movement gave us a hint about the power within the mind. However, the basis for this developed much earlier.

The New Thought Movement began in the late 1800s and emphasized that Infinite Intelligence dwells within each person (oneness of the human race with Infinite Intelligence or God), the highest spiritual principle is unconditional love, the power of thought, especially in healing, and the law of attraction, basically an optimistic scheme of life. From Phineas Quimby (1802-1866), Napoleon Hill (*Think and Grow Rich*), Wallace Wattles, Prentice Mulford (1834-1891, featured in *The Secret*) to William Walker Atkinson (1862-1932) who published *Thought Vibration or the Law of Attraction in the Thought World* in 1906. The next year Bruce MacLelland published *Prosperity Through Thought Force* in which he summarized the "Law of Attraction" as a New Thought principle, stating "You are what you think, not what you think you are." Out of this grew the Unity Church, Religious Science, and Church of Divine Science with many of its churches and community centers led by women, from the late 1880s to today.[3]

Ralph Waldo Trine's *In Tune with the Infinite* was a lifesaver at a difficult time in my life, and I can't remember how I came to possess this tattered old book from 1897. Trine (1866 – 1958), a philosopher, mystic, teacher and author of many books was one of the early mentors of the New Thought Movement.[4] Henry Ford attributed his success to having read this book and distributed copies to other industrialists. Trine wrote, "Within yourself lies the cause of whatever enters into your life. To come into the full realization of your own awakened interior powers is to be able to condition your life in exact accord with what you would have it." Trine was influenced by his namesake, Ralph Waldo Emerson, who wrote, "What lies behind us and what lies before us are tiny matters compared to what lies within us." I found it interesting that this was quoted recently by 'Hoetchner' on the popular CBS® TV series, "Criminal Minds."

I read *Think and Grow Rich* (Napoleon Hill, 1883-1970) in the '70s, but James Allen's (British philosopher, 1864 -1912) book *As a Man Thinketh* published in 1906 was the 'pioneer' self-help book. I loved Marc Allen's edited version, *As You Think*, published in 1987. He calls the material 'self-empowerment' and pointed out that the principles so clearly are universal, applying to everyone regardless of sex, age, race, beliefs, social standing, or education.[5] James Allen's original book, one of nineteen he wrote in twelve years, several published by his wife posthumously, took its title from the Bible quote, "As a man thinketh in his heart, so is he." This seems very similar to what the Buddha (563 BCE – 483 BCE) taught: "All that we are is the result of what we have thought."

When Maharishi Mahesh Yogi brought Transcendental Meditation to the U.S. it seemed the 'West' was ready for this venture into eastern philosophies; the Bhagavad-Gita became a choice of students. The training of teachers and initiations of beginners spread like wildfire. Discussions about consciousness were on every campus. When the music stars, the 'Beatles' travelled to Maharishi's ashram in India to be initiated, the whole world heard about it. The research on the beneficial effects of meditation, published by respected professors at Maharishi International University in Fairfield, Iowa, were noticed and reviewed by scientists at prestigious universities such as Harvard. And, the discussion of the brain wave levels of alpha, beta and delta became 'mainstream.' Maharishi said, "the means gather around *sattva* (purity of desire). All of Nature is vibrant with anticipation of the Descent of Heaven on Earth; Nature will therefore be anxious to support those who desire* it."

Maharishi postulated that regularly practicing TM enabled one to get in touch with a quantum energy field that connects all things. When a group of meditators was large enough, he claimed their collective meditations caused "Super Radiance," a term also used in physics to describe the coherence of laser light. During TM, the theory went, the minds of meditators all become tuned to the same frequency, and this coherent frequency begins to order the disordered frequencies around it. Resolution of individual internal conflict leads to resolution of global conflict, states Lynne McTaggart. ". . . the theory rests entirely on the premise that meditation has a threshold effect. If 1 percent of the population of a particular area practices TM, she claims, or the square root of 1 percent of the population practices the TM-Siddhi program (a more advanced type of meditation) conflict of any variety – the rate of murders, crime, drug abuse, even traffic accidents – goes down."[6] Numerous studies have demonstrated this positive effect in many areas and types of focus.

R. Lataine Townsend

I heard about TM in the '70s from friends who had been initiated in college and soon found that the course was being offered in my small town in Colorado. So began my own 'journey' into the world of consciousness and eastern religions. Though evolving over the years as I delved more deeply into the consciousness theme, TM has been a wonderful anchor in the seemingly uncontrollable sea that life can become. Dr. Deepak Chopra, noted endocrinologist, was associated with the TM Movement in the 1980s, facilitating their Siddhis and Ayurvedic programs. He began publishing books that helped bridge the gap that seemed to have opened between the human potential scene and western religions - a gap many of us had fallen into. We couldn't quite yet see ourselves as self-help author, Tony Robbins saw us (especially the fire-walking part!), but could no longer agree with much that was happening in the Christian religions either.

As the self-help category grew, the space programs reached further into space, the Chernobyl nuclear reactor blew up, there was the Harmonic Convergence of 1987, which may have been the first time I heard of the Mayans and their calendar, and crop circles became more numerous. Wayne Dyer filled that psychological/motivational/ spiritual opening that so many people's hearts were aching for, with books like *You'll See It When You Believe It*.[7] These led some, like me, to go back and read the work of earlier writers such as Catherine Ponder, Louise L. Hay, Talbot Mundy, Gina Cerminara and, one of my all-time favorites, *The Door of Everything* by Ruby Nelson which, in one section, reinterpreted the message of love that Jesus brought.[8] It left me feeling loved by God, The All-That-Is, as religion never had. Gregg Braden began exploring and writing in the '90s and I have been a 'groupie' since the first book of his that I read. His early book, *Awakening to Zero Point*[9] seemed ahead of its time, yet somehow I had missed it until recently.

Once the 'levee' was breached, books of inspiration, metaphysics, and Native American spirituality, to name just a few categories, by the thousands poured out. Self-actualization became the new 'catch phrase.' People who had gifts, like the medical intuitive, Caroline Myss, now felt comfortable to share their 'knowing.' Sensitives and those channeling discarnate entities, like *Lazaris, Theo, Seth* channeled by Jane Roberts, *Kryon*, through Lee Carroll, and, of course, *Raphael*, who gave Ken Carey *The Starseed Transmissions, Return of the Bird Tribes*, and *Vision*, were published and added greatly to our broadening base of inspirational knowledge.

Jean Houston, Ph.D., author of 26 books, researcher of consciousness, spirituality and ritual processes, wrote the Introduction to *Vision*, in which she praised "its sense of being in touch with the larger Pattern that was trying to enter into time." She said Ken's books give us "our opportunity for co-creation with God at this most critical turning point in human history." (1985) She gave an excellent description of channeling: "Channeling begins and ends in awareness, awareness of the enormous amounts of information coming to us all of the time. There is much to suggest that in pre-historic times we were much more generally aware of different cadences of information than we are now. . ." "Sensitives like Ken Carey, especially if they live in quiet natural settings are not only more responsive to their environment, but are always enhancing their ability to sense or channel depth realms of Being as well." She gave examples of this type of knowledge from the 6th century BCE, Plato, Jesus, and Sufi Islamic mystics who refer to the *alam al mithal*, the *mundus imaginalis*, an intermediate universe that is thought to be as real as the sensory world, but that can only be experienced by those who exercise their psychospiritual senses (which is very close if not identical to channeling) . . ." (For many of us) "our 'cerebral reducing valves' protect us from this avalanche of cosmic knowing. "What we experience and label as . . . gods or guides or even angels . . . what they really are and where they are coming from God Only Knows. Be that as it may, the universe bleeds through and we are diaphanous to its rhythms and knowings. Seen from this perspective, channeled information is a by-product of this simultaneous-everywhere-matrix of reality. . ." Dr. Houston continues, "Nostradamus channeled pictures . . . of times to come. Throughout the shamanic cultures, (we see) the shaman consciously alters his attention on the spectrum of consciousness in order to access levels of knowing that are not usually available to ordinary states of consciousness. . ." ". . . the nervous system and the brain have to be re-educated in order to open the doors of perception on the strange and beautiful country of channeled knowing. Otherwise, one gets the great garbage heap of the unconscious . . ." "They (those who channel)* speak to a vision and a gnosis (meaning knowledge through direct experience or personal revelation)* deeper than any culture (and) more universal than any theology."¹⁰ The Mayan shamans were very familiar with this type of connection to the *Otherworld*.

Vision begins with a Creation Story that is as beautiful as anything I have ever read – and very similar to those of some of the indigenous people, especially the Mayans. He relates that this is "a pregnant moment, a pregnant quarter century." (1985-2012)* "Throughout the universe all conscious beings watch. If the human family chooses to ask for guidance, the coming Awakening will be the most beautiful spectacle ever to grace time. The choice is human." "You have incarnated at this time specifically to help ease this transition." "The history of the next two decades is a blank slate, waiting for those who will welcome my Spirit into their lives. Human history has been defined in fear. . . . I call upon you who have been drawn to these words and who are able to understand them to use your intelligence, ingenuity, all the cultural tools of expression that are available to you for the purpose of helping me ease this greatest of all human transitions." *Vision* ends with this promise: "My consciousness is the gift that I offer to all children, all women, all men of all races, tribes and nations who choose to dedicate themselves to LIVES OF LOVE."¹¹

Consciousness Defined

*L*yn Birkbeck gives a definition of 'consciousness' from Dictionary.com as: "the state of being conscious; awareness of one's own existence, sensations, thoughts, surroundings, etc." From his point of view, "Everything is waves. Thoughts are waves. Your thoughts are an expression of your consciousness. Your consciousness is a succession of waves. The waves go out and contribute to the creation of reality. So your state of consciousness has something to do with the state of the world." "... both astrology and quantum physics are essentially spiritual in their outlooks, as it is understood that life energy streams eternally from Source and takes form as living things – which includes you and I – and then returns to stream forth again, ad infinitum." "... witnessed in astrology as the endless movement of heavenly bodies, and in quantum physics as the endless emission and interchange of waves and sub-atomic particles." "Let it be said that 'spiritual view' includes any approach to life that is guided by higher principles or higher beings, and not exclusively the planets and their waves." "Most of all . . . that you as an individual have the power to influence the whole, in a small or great way – because the whole is made up of millions of individual wills and minds. . ."

He writes, "a 'second coming' is more likely to be that growing group of individuals who are giving expression to the New Super-Paradigm. When it reaches critical mass there will be a quantum leap in collective awareness, and then a shift in what our reality itself is." "... each one of us is like a little drop in the River of Life, each making our minute but significant contribution to its state and direction." "... consciously or unconsciously, we have arrived at a time of reckoning." "... or it could be the arrival or intensification of some unknown energy within the Earth's biosphere;

something akin to whatever it is that so mysteriously creates crop circles."[12] He goes on to delineate three possible scenarios that are dependent on one's inner conflict or sense of peace that I find very compelling, and it inspires me to try to be in that second category of inner peace!

Gregory Sams in *Sun of gOd* asks, "Is our particular brand of consciousness a higher incarnation of the divine light of the cosmos than that of other species? Perhaps it is, and our higher level of intelligence is an indicator of just how much light we are capable of *channeling*, whether we choose to use the faculty or not.[13] In *The Orb Project* (Ledwith & Heinemann) we see that the "realm of the paranormal and the mystical are not really different realities." We need "to open up to the immense possibilities that new realms are beginning to unfold about our . . . significance in the cosmos." "(The) fear is . . . not our weakness, limitations, and inability to cope, but the greatness that we suspect lies hidden just beneath the surface of what we are. Most of the great thinkers . . . have known how fearful we are of the responsibility that comes with realizing we create our own destiny." Orb photos taken at a spiritual retreat of healer, Ron Roth, featuring a speaker from the Oneness University in India, show that a few Spirit beings were apparent in the beginning, but by the end of the talk, there were crowds of the Spirit beings (orbs), at least many who chose to make their presence known.[14]

Author and TV show host, Dr. Meg Blackburn Losey, writes in *The Mystery of 2012*, "Our consciousness is separate from our thought process . . . (consciousness) knows no boundaries and is *always* aware of what is happening in our world and beyond, in other dimensional reality. In fact, our consciousness exists fully within all levels of reality, all of the time; and it is doing its best to get us to listen to the clues and signals that it picks up in all of creation." "The true meaning of the ancient prophecies relating to the year 2012 is that we are going to take a major leap in human evolution . . . as we enter into a shift of ages. . ." "A change of ages requires a change in consciousness – a different way of existing altogether." "Our brains and our DNA begin to reawaken. That includes seeing through the etheric veils and into other dimensions, spontaneous gifts of telepathy, remote viewing, full sensory knowing of things to come and of other's thoughts and motivations. This is what I call our Seventh Sense, otherwise known as our multidimensional, universal sense," writes Dr. Losey. "As our electromagnetic fields change and our electrical patterning is altered . . . our harmonic frequencies change. They become higher, more similar to the harmonics of our source." "The end of the Mayan calendar . . . did not mean that we would disappear from our planet; rather, we could step into a new form of reality and the age of ignorance . . . would come to an end." She also suggests, "We must acknowledge our self-perfection. We are and always have been one with Source." "We must acknowledge our true value . . . accept our power . . . and take those into (the) world." Lastly, we must "love ourselves and touch everyone we encounter with love."[15] Excellent advice, I say.

Gregory Sams defines consciousness as "the power behind our mind and being, essentially an invisible phenomenon." As he says, "There is no accepted standard definition. . ." "Mystics and philosophers have devoted entire books to it or summed it up with two words: consciousness is." After citing six 'definitions' from *The Oxford English Dictionary*, he comes to the simplest: 'with knowing.' He writes, "Perhaps there is but one universal field of consciousness, and what we perceive as our own is simply the unique interface that our mind and personal spirit have with it. We can at times expand our consciousness to embrace the infinite Universe, or live a small-minded life focused entirely upon the props surrounding our physical existence," Sams concludes. Life force energy, called *Chi* in the Asian cultures, *Prana* in India, is similar to that which the Maya termed *Ik*. Holy Spirit may be the comparable term for Christians." ". . . the divine spirit is always perceived to be something with which our consciousness is fundamentally intertwined, or 'All is One.'"[16]

"Pure consciousness is Life before* it comes into manifestation," writes Eckhart Tolle in *Stillness Speaks*, "and that Life looks at the world of form through 'your' eyes because consciousness is who you are. When you recognize yourself at That, then you recognize yourself in everything." "Through 'you,' formless consciousness has become aware of itself." He further states, "Wisdom is not a product of thought. The deep *knowing* that is wisdom arises through the simple act of giving someone or something your full attention. Attention is primordial intelligence, consciousness itself."[17] In *A New Earth* Tolle continues his explanation, "Consciousness . . . is the unmanifested, the eternal." "Consciousness (with a capital 'C') itself is timeless and therefore does not evolve." "When consciousness becomes the manifested universe, it appears to be subject to time and to undergo an evolutionary process." "(It)* is the intelligence, the organizing principle behind the arising of form. Consciousness has been preparing forms for millions of years so that it can express itself through them in the manifested. Although the unmanifested realm of pure consciousness could be considered another dimension, it is not separate from this dimension of form. Form and formlessness interpenetrate. The unmanifested flows into this dimension as awareness, inner space, or Presence. Consciousness incarnates into the manifested dimension, that is to say, it becomes form." "When the brain gets damaged . . . it means consciousness can no longer use that form to enter this dimension. You cannot lose consciousness because it is, in essence, who you are."[18]

"Consciousness . . . arises whenever a reflection in time occurs. That means something reflects from points in the present or the future and (points in) the past, or even both past and present, or future and present. What reflects depends of the form of consciousness," explains Dr. Fred Alan Wolfe. "If we're talking about primal reflections from the beginning and ending of time, then the reflection produces a conscious and cosmic soul. I would call this the one Soul that inhabits each and every being."[19] Dr. Wolfe is a scientist, but notes in the Introduction that "People from various scientific, religious, and philosophical disciplines have begun building bridges between science, spirituality, shamanism, ancient magical practices, metaphysics, and the functioning of the human body, among other areas."

Another viewpoint comes from the ancient roots of Melkezedekian theology and their interpretation of world ages. "Their tradition speaks of *Missiayah,* (we are familiar with the word *Messiah*) . . . (which) in its ancient meaning was not a person; it was an energy field, a consciousness, that every 2000 years entered the earth plane." "And if the earth was ready to receive this energy of expanded consciousness, it would 'alight' . . . if the earth was not ready (to embrace it), it would come again . . . 2000 years later."[20] The Essenes, living in ascetic spiritual communities around 2000 years ago, were preparing the way for this new consciousness, and many believe Jesus of Nazareth studied many years with the Essenes. This is very close to the 2125 years of each Zodiacal age and, from this ancient point of view, will come again soon. Or as Daniel Pinchbeck writes, ". . . intensified consciousness that Christ (Jesus) brought to the Earth was too much for (hu)mankind – save for a few – up until the present day." He "realized that would be the case." "World avatars are frequency transducers who step up the voltage of mind." ". . . parables in the New Testament and in the *Gospel of Thomas* – found, along with other Gnostic scriptures . . . could be considered devices for storing and transmitting higher energies." "In the Gospel of Thomas, Christ . . . declares, 'If you bring forth what is within you, what you bring forth will save you . . .'"[21] In the chapter "Gnosis: The Not-So-Secret History of Jesus" (*Toward 2012*, edited by Pinchbeck and Jordan) Jonathan Phillips reports that *The Gospel of Thomas* was the earliest gospel, written in 40 AD, and is known as 'the secret sayings gospel.' He quotes from it: "Whoever discovers the interpretation of these sayings will not taste death." "Heaven is inside and outside you. When you know yourselves, then you will be known, and you will understand that you are children of the living father." Phillips adds a bit of history: "The Romans (who overthrew Jerusalem and its Second Temple in AD 70) considered secret or hidden societies dangerous hotbeds of rebellion and Christians, with their radical messianic hero figure, found themselves at the top of this list." "Initiates spread far and wide. . ." He relates, "To conclude this chapter of our journey,* I'd like to say that I've been absolutely amazed by how many people are awakening to a greater vision of themselves and the cosmos, whether through spontaneous opening or engaging in serious spiritual endeavors. Mass transformation of human consciousness seems to be increasing exponentially all around us as record numbers of seekers practice the techniques of yoga, Reiki, Tai Chi, meditation, and much more." "While we embark on this noble journey . . . we need to remember the quote from Thomas above, as well as the last part, which is, 'If you do not bring forth what is within you, what you do not bring forth will destroy you.'"[22]

The Oneness Movement

Another strong consciousness movement began with Sri Kalki Bhagavan when he announced that the Hindu Golden Age would begin in 2012, much to the displeasure of the Brahman spiritual establishment. He and his wife, Amma, established the Golden Age Foundation and the Oneness University in India and today more than 300 million people have experienced the Oneness blessing or *Deeksha*. What began as a simple school for children, who began to experience the higher states of consciousness which they called the 'Golden Ball of Divine Grace,' became a spiritual center to raise the levels of human consciousness and rapidly outgrew its space. After 1996 the Oneness Movement spread to other continents, mostly by word-of-mouth. In 2003 at Nemam, India 100,000 people received the Oneness Blessings from Sri AmmaBhagavan and experienced amazing transformations and various states of consciousness. There are 60,000 active Oneness Deeksha Givers worldwide.[23] "The Oneness Deeksha is a major leap in humanity's spiritual journey, for it directly affects the neurobiology of the brain and results in a direct experience of Consciousness," states their website. "Since the phenomenon is primarily experiential in nature, throughout its unfolding it has thoroughly been embracing every faith, belief system or spiritual tradition . . . awakening (each person) to their own truths, their personal Divinity and their own sacred self." Having outgrown the original spaces, the Oneness Movement established the Oneness University with the Oneness Temple and World Oneness Center on several thousand acres in the Chittoor district in South India.[24]

Arjuna Ardagh, who had an interest in spiritual awakening beginning in his teens, graduated from Cambridge University with a master's degree in literature, and studied with a number of great spiritual teachers including

H.W.L. Poonjaji, a devotee of the great sage Ramana Maharishi in India. In 2005 Ardagh was invited to the Oneness University and now integrates the Oneness Blessing into his work at his Living Essence Foundation in Nevada City, California, which is dedicated to the awakening of consciousness within the context of ordinary life.[25] He wrote *Awakening Into Oneness,* (as well as *Leap Before you Look* and five other books) in which he says the blessing is a form of coherence (Note how that word keeps coming up*) that can be transferred from one individual to another. Ardagh visited the villages in which the original blessing was taught and reported that alcohol consumption there was reduced by 80% and compared the benefits in the communities to the *Maharishi Effect,* scientifically documented by the TM scientists in 1993.[26]

Lawrence E. Joseph reports that Sri Bhagavan ties his 2012 prediction to the transit of Venus (crosses the face of the Sun) that happens twice each century eight years apart. The last transit was June 8, 2004 and will occur again on June 6, 2012. The "Mayans noted this also and believed that Venus embodied their supreme deity of goodness, the feathered serpent, known as *Kukulcan,*" writes Joseph. "In Vedic mythology Venus is called *Shukra* . . . which, in Hindu numerology, governs the number six. The next transit will occur on 6/6/12 (6+6)."[27] Interesting coincidence! I have received the Oneness Blessing two times; I did not have a mystical experience. However, I believe my focus, clarity and feelings of connection with All-That-Is have increased since and I plan to seek this powerful blessing again. As you can see I have a rather eclectic approach to spiritual practice.

A Wealth of Encouragement

There are so many books, seminars, DVDs, classes and articles 'out there' on consciousness it would take pages to even begin to list them. In my small library alone are several by Thich Nhat Hanh, His Holiness the Dalai Lama, Kahlil Gibran, Hafiz, four translated books of Jelaluddin Rumi's work, Eckhart Tolle, seven by Neale Donald Walsch, books by Joseph Campbell, Jean Houston, Peter Matthiesson, Fritjof Capra, Stephen Hawking (on physics), Raymond A. Moody (life after death), James F. Twyman, Clarissa Pinkola Estés, Ph.D., Mike Dooley (books and DVD), Michael Bernard Beckwith and others featured in the movie/DVD *The Secret*, as well as Rhonda Byrne's two books, *The Secret* and *The Power*, and the movie/DVD *What the Bleep Do We (K)now?!!*

The Handbook to Higher Consciousness: The Science of Happiness by Ken Keyes, Jr. and Richard Bach's books *Jonathan Livingston Seagull* and *Illusions* fed my appetite on this subject as the '70s turned to the '80s. I gained a lot of insight from a week-long course at the Ken Keyes Center in Coos Bay, Oregon in 1983. Three other early writers were Catherine Ponder (*The Dynamic Laws of Prosperity*), Stuart Wilde (*The Force*) and Sue Sikking whose book title says it all: *God Always Says Yes! . . . to the power of choice in your mind.*

Then there is Echo Bodine, Martha Beck, Lama Surya Das, Osho, Masaru Emoto, Shakti Gawain, Joan Borysenko, Ph.D., Debbie Ford, Arielle Ford (*Hot Chocolate for the Mystical Soul*), Marianne Williamson, Marci Shimoff, Talbot Mundy, Harry Palmer (the Avatar Materials), and Gary Zukav; all names you will recognize if you have been 'into' this 'scene' very long; if not, I heartily recommend them. And this doesn't even 'scratch the surface' of what is on the market about consciousness. Even

2013

Dan Brown's fiction (*The Da Vinci Code* and *Angels and Demons*) opened a wealth of new ideas for all of us.

I have mentioned previously works by Alberto Villoldo, Ph.D., Gregg Braden, James Redfield, Wayne Dyer, Deepak Chopra, M.D. and many others that have informed us along our 'journey.' Even a health related book, *The Menopause Thyroid Solution*, by Mary J. Shomon has a chapter on "The Role of Mind, Body and Spirit," quoting Indira Gandhi, "You must learn to be still in the midst of activity and to be vibrantly alive in repose." Shomon writes, "Noted guided imagery therapist Bellaruth Naparstek explains that the brain doesn't know the difference between something you actually see and something you imagine. So your body can respond as strongly to an image or a thought as to the real thing."[28] These writers may be touching on similar subject matter, but they are each addressing this theme from their unique perspective and experiences. The *One Spirit*[29] book club is a great resource for books of inspiration, health, spirituality, and consciousness, just to name a few categories.

Dr, John J. Liptak prefers a more positive interpretation of the Mayan prophecy and our evolving consciousness, stating, "human beings will become more aware of nature and the impact that we have on . . . the planet . . ." (They) "will begin to develop a higher intelligence, greater consciousness, and greater willingness to lead lives of service." ". . . the sun will intersect with the Milky Way causing an eclipse that will allow everyone to see the Mayan 'Tree of Life'. . . (and) a new and fifth dawn of enlightenment will evolve."[30] As we recall, the Tree of Life was symbolized by the Mayans in the shape of a cross. (It was the center) "through which all vital energy from within the earth traveled, and through which all . . . from the gods traveled." ". . . people experienced psycho-spiritual energy from the Center, East, West, North and South, . . . each one produc(ing) a very different type of energy." "Mayan scholars believe the current values and beliefs . . . what the Mayans call *Koyanisqoatsi* or 'world out of balance' will expire* (on 12-21-12) and a new stage of spiritual growth begin."[31] FYI: There will be a total solar eclipse on November 13-14, 2012 visible in eastern Australia, the south Pacific, and a partial in Chile.

Quoting some selected esoteric and mystical passages from *The Starseed Transmissions* by Ken Carey may shed some light on the importance of the time period we are approaching. The book begins with, "I am come from the Presence where there is no time but the eternal now. I come with a message that will prove vital to you. . ." "The total cessation of movement . . . is a micro-interval of non-time – the same that occurs many times each second as the atoms of the physical world vibrate back and forth. This is an opening into the *nagual*,* a doorway into the Presence from which all Life-energy springs." "Existing within this ripple of non-time will be the focused conscious attention of the Creator. Indeed no single conceptual structure is capable of conveying the enormity of what is soon to take place." "To those . . . who have tuned themselves in to the will of (the Creator), the coming interval of non-time will literally expand into eternity. . ." "You will know . . . that you* are the bridge between Spirit and Matter, between Creator and Creation . . ." "There is a new vibrational pattern descending upon your planet. Two worlds of consciousness will begin to form ever more distinctly; the world of Love and Life, and the world of fear and death . . ." "The polarization will continue to intensify. Except in a few isolated pockets where the new energies will not enter until later, the momentum of positive change is being felt everywhere."[32] (Sounds good to me!)

In Transcendental Meditation the 'objective' is to attain coherence between the left and right hemispheres of the brain, thus putting the body in the optimal 'space' for healing, creativity and rest. This 'bringing together' of various types of duality (such as right/left, male/female, black/white) seemed to be the aspiration of ancient as well as modern shaman/teachers. Mesoamerican myths contained the polarities of Upperworld and Underworld, Mother Earth and Father Sky, life and death, yet they were inseparable like the yin/yang symbol. (See Illustration #10 - the Hunab Ku symbol compared to Yin-yang, images from dreamstime.com and wikipedia)

Hunab Ku and Yin/Yang

Physicists are trying to find *The Unified Field* that 'brings all theories together; spiritual teachers keep saying there is no black or white, Christian or Muslim, East or West, we are all One. We are all children of Mother Earth and Father Sky, both made by the One Creator, the All-That-Is, or the Great Mystery, as some indigenous people refer to God.

Barbara Hand Clow believes we must 'see' from the galactic perspective to peer into the true nature of reality as the Mayans predicted we would at the end of their calendar. She says, "Everything that exists comes first from consciousness, since the material world emanates from creative intentions; thus, consciousness drives evolution." She predicts that beginning in 2011 "love will be the greatest force on Earth." She also speaks of the Uranus/Pluto square, which we examined under "Astrology and Astronomy" with the Mother Wave calculations explained by Lyn Birkbeck. Clow writes, ". . . the exact squares are during 2012 on June 24 and September 19" which predicts "the enlightenment energy discovered by the Children of Love will manifest as a global force in 2011 and during 2012, no one who lives on the Earth will be able to resist melting into enlightenment."[33] In light of the quantum physics research that nothing travels faster than light (except thought) I love Ms. Clow's statement, "Love travels faster than thought!" I just saw a review of a book by Sonia Choquette, *Traveling at the Speed of Love*, destined to be another aid in our 'journey' toward greater awareness.

Kryon relates, through Lee Carroll, "It has been coming, and you have had the channelings that said so. It is the Great Shift and it begins . . . the preparation for what the Mayans told you would happen as the Earth shifts into the year 2012. This magic year is only a signpost . . . that says you are moving into a new energy that was foretold by the angels. It is going to be different. For those of you who don't like change, it's going to be fearful. All these things I bring are positive, filled with light, (and) even within change . . . the slow development of peace on Earth."[34]

The Science of Consciousness

Hal Zina Bennett's book, *The Lens of Perception*, is all about consciousness, but I particularly like this statement: "In spite of the great crises of our times – global warming, 9/11, war in the Middle East – it is still my belief that we are on the threshold of vast positive changes in the world order, based on a change of consciousness.* Even while others warn of the apocalypse, there is growing awareness of the fact that a change of mind is far more powerful than all the weapons of war, and it is a power that we are learning to harness for the good of our whole planet. . ." "We must recognize that what we project onto the world from our lens of perception can result in war and destruction or be transformed, so that it helps to build peace and love." "Each one of us should recognize the role we play in expanding of these matters." In his Epilogue Bennett quotes Elisabet Sahtouris, "Our biggest job is to change our whole way of thinking to a larger perspective, to recognize ourselves as a body of humanity embedded in, and with much to learn from, our living parent planet, which is all we have to sustain us."[35] I think the Mayans would agree with that statement!

Another point of view comes from Layne Redmond (*When the Drummers were Women*): "The ancients studied the stars and the earth with humility and a desire to participate as fully as possible in the rhythms of the universe." "(The) change from calendars based on nature to calendars designed to serve the purposes of those in power was one of the earliest rejections of the truth of direct experience." We looked at the pineal gland in relation to the geomagnetic field but Ms. Redmond examines it from the perspective of biorhythms: "Research . . . suggests that the pineal gland is indeed a kind of light sensor. The sun and, to a lesser extent, the moon, pour energy into the biological environment in the form of light . . . it appears that the pineal gland may be the jewel of our biological clockworks, keeping us in sync with environmental time and influencing the physical and emotional rhythms of the body." ". . . modern science is coming to the same conclusions the ancients knew from immediate experience: that life is inexorably rhythmic." "As the body takes shape in the womb, consciousness begins . . . detected as brain waves." EEGs (electroencephalograph) can measure the number of energy waves per second pulsing through the brain. Active, waking attention ranges from 14 to 21 (some say 28 or up to 40, according to Lynne McTaggart) cycles per second is called **beta**; a relaxed, meditative state, called **alpha** waves are 7 to 14 per second; semi-conscious, threshold of sleep waves, 4 to 7 cycles per second are called **theta**; and the slowest, 1 to 4 waves per second, **delta**, are found in deep sleep, unconscious and *in fetuses*. The alpha range is found in animals and "correlates to the electromagnetic field of the earth." (We will see later that some think this rate is rising.)

"So when the Tantric Yogi sought to bring herself into alignment with cosmic vibrations, she aspired to the alpha state, which is in fact the basic rhythm of nature. This is the process referred to by meditational adept Swami Rama as 'stilling the conscious mind and bringing forth the unconscious.'" Ms. Redmond also shows how people throughout history have sought various means to connect with this 'tempo' of the earth, beginning even before the ancient labyrinth dancing in Crete. "Guided by drumbeats, these sacred drummers (priestesses of the goddess) could alter their consciousness at will, traveling through the three worlds of the goddess: the heavens, the earth and the underworld."[36] This slowing to the 'beat' of Mother Earth is the goal of drumming, Native American ritual dancing, meditation, the relaxation technique and other methods from around the world.

Vianna Stibal relates, in her book, *Theta Healing*, "Scientists have discovered that certain brain frequencies (particularly in the Alpha, Theta, and Theta-Gamma states) have been found to (1) alleviate stress and promote reduction in anxiety, (2) facilitate deep physical relaxation and mental clarity, (3) increase verbal ability and performance, (4) synchronize both hemispheres of the brain, (5) invoke vivid spontaneous mental imagery and imaginative creative thinking, and (6) reduce pain, promote euphoria and stimulate endorphin release."[37] Just being in nature, in the mountains, beside the ocean (two of my personal favorites) or hiking in the desert, can bring us closer to this connection that is so healing to our soul and psyche, as well as our bodies.

When the former Czech president, Václav Havel, in 2007 presented the *Vision 97 Award* to Stanislav Grof, M.D., renowned researcher into nonordinary states of consciousness, the recipient spoke of his early work with LSD and various frequencies of oscillating light, in which he had a mystical experience. It was so profound that he devoted his career to exploring the transformative and evolutionary potential of these states of consciousness. His work with people undergoing spontaneous 'psychospiritual crises,' and 'nonordinary' states of consciousness changed his understanding of consciousness, the human psyche, and the nature of reality. Dr. Grof came to the conclusion that various phenomena, "such as statistically highly improbable meaningful coincidences (Jung's 'synchronicities'), shows the inevitability of a radical revision of thinking in psychology and psychiatry." "New observations show that consciousness is **not** . . . a product of complex neurophysiological processes in the brain, but a fundamental primary attribute of existence, as is described in the great spiritual philosophies of the East."

"The human psyche is an integral part of this cosmic matrix . . . and as C.G. Jung said, permeates all of existence." "Feelings of oneness with the universe and its creative principle lead to identification with all sentient beings and bring a sense of awe, wonder, love, compassion, and inner peace. Spirituality that results from this process is universal, all-encompassing, transcending all organized religions; it resembles the attitude about the Cosmos found in the mystics of all ages . . . extremely authentic and convincing, because it is based on deep personal experience." "This is based on an . . . understanding that any boundaries in the Cosmos are relative and arbitrary, and that each of us is, in the last analysis, identical and commeasurable with the entire fabric of existence," writes Grof.[38]

The Pleiadians taught, through Barbara Marciniak, "As energy increases on the planet, blocks in your physical, emotional, and spiritual bodies are magnified. Unexpressed feelings and ideas create obstacles to the flow of energy. Wherever you have a difficulty or prejudice – you can trust that the magnifying glass will be put (on) it. Everything is intensifying in order to teach (you) . . . about maintaining a clarity of purpose and intent." "Actually, *you* invite all the players in your life, and you, as director, cast the parts and run the show. If you are now finally tired of your script, remember, *you* write it! Blame and victimhood are the ultimate traps to insure a slate of disempowerment. All of reality is connected and is seeking a healing of union." "There is great humor in highly evolved energies, especially those who work with the love frequency. It can be recognized as a trademark . . . for laughter is a key to freedom. There is plenty of room for joy in all of existence. We encourage you to operate out of your feeling center – your solar plexus . . . this is where you hold power in your body." "Everything occurs in the now. This very moment of existence – where you are – is truly the ongoing, spontaneous, significant moment. It is, over and over again, where you can find yourself*."[39] This reminds me of the work of Eckhart Tolle, a must-read for anyone interested in self-improvement or evolving consciousness.

The Power of Intention

The mention of 'intent' brings us to the research of Lynne McTaggart, who is probably best known for her books, *The Field* and *The Intention Experiment*, bringing science and consciousness together in a unique way. In the Preface of the second book, she describes the "extraordinary quantum field generated by the endless passing back and forth of energy between all subatomic particles. The existence of this Field implies that all matter in the universe is connected on the subatomic level though a constant dance of quantum energy exchange." ". . . on the most basic level, each one of us is also a packet of pulsating energy constantly interacting with this vast energy sea." ". . . experiments conducted by scientists suggested that consciousness is . . . outside the confines of the body – a highly ordered energy with the capacity to change physical matter. The mind-over-matter power even seemed to traverse time and space."[40]

McTaggart writes in *The Intention Experiment*, "*The Field* created a picture of an interconnected universe and a scientific explanation for many of the most profound human mysteries, from . . . spiritual healing to extrasensory perception . . ." (It) "suggested . . . that directed thought had some sort of central participatory role in creating reality." "Targeting your thoughts, or intentionality, appeared to produce an energy potent enough to change physical reality." The huge response she received from the book caused her to wonder how some of this could be applied to personal reality and even the planet; so she began to research 'intention.' She turned to science, as well as those "who had managed to master intention – spiritual healers, Buddhist monks, Qigong masters and shamans." The research included that gathered by the Transcendental Meditation organization. She then enlisted the help of prominent scientists from Princeton and the

International Institute of Biophysics in Germany, all 'sticklers for strict scientific method.' As Ms. McTaggart says, "It is important to recognize that the conclusions arrived at . . . represent the fruits of frontier science. . ." "To be a true explorer in science . . . is to be unafraid to propose the unthinkable."

The Intention Experiment explains one of the best mind-over-matter experiments ever done. The "evidence suggests that human thoughts and intentions are an actual physical 'something' with the astonishing power to change our world." This central idea, that consciousness affects matter, lies at the heart of an irreconcilable difference between . . . classical physics . . . and that of quantum physics. . . (in which) subatomic matter appeared to be involved in a continual exchange of information, causing continual refinement and subtle alteration. (And that) the universe was . . . a single organism of interconnected energy fields in a continuous state of becoming."[41]

We must understand this basis of our physical world to understand how and why consciousness is SO powerful. McTaggart explains, "Every subatomic particle is NOT* a solid and stable thing, but exists simply as a potential of any one of its future selves, or the sum of all probabilities . . . an ephemeral prospect of seemingly infinite options." "At the quantum level, reality resembled unset Jell-O." "Once in contact, particles retained an eerie remote hold over each other. The actions of one – for instance, magnetic orientation – of one particle instantaneously influenced the other, no matter how far they were separated." ". . . these little packets of vibrating energy constantly traded back and forth to each other via 'virtual particles' like ongoing passes in a game of basketball . . . that gave rise to an unfathomably large basic layer of energy in the universe." (It was) "a continual exchange of information, resembl(ing) a vast network . . . with all its component parts constantly on the phone." "It implies that reality is not fixed, but fluid, or mutable, and hence possibly open to influence – that observation – the very involvement of consciousness gets the Jell-O to set."[42]

"In classical physics, a field is a region of influence, in which two or more points are connected by a force, like gravity or electromagnetism. However, in the world of the quantum particle, fields are created by exchanges of energy – always being redistributed in a dynamic pattern."

(Heisenberg's uncertainly principle)* McTaggart explains, (These) "little knots of energy (from the quantum wave to particle transition*) briefly emerge and disappear back into the underlying energy field. These back and forth passes, (give) rise to an extraordinarily large ground state of energy known collectively as the Zero Point Field – called so because even at the temperature of absolute zero, when all matter theoretically should stop moving, these tiny fluctuations are still detectable – never come to rest and carry on this little energy tango." "Richard Feynman (physicist) once remarked that the energy in a cubic meter of space was enough to boil all the oceans of the world." Multimillion dollar projects are attempting to develop this Zero Point energy for space travel.

McTaggart relates, "Benni Reznik (Israeli physicist) theorized that "if all matter in the universe was interacting with the Zero Point Field, it meant, that all matter was interconnected and potentially entangled throughout the cosmos through quantum waves" which offers an explanation "for why signals being generated by the power of thought can be picked up by someone else miles away." The latest in physics research "suggests that the observer effect occurs not simply in the world of the quantum particle but also in the world of the everyday. Our observation of every component of our world may help to determine its final state, suggesting that we are likely to be influencing every . . . thing we see around us."[43]

She cites many types of consciousness research projects, but her own work with 'intention' research is very thorough and 'leading edge.' She began with research on healers and found that while a person is at rest their electrostatic energy reading on an EEG is 10 – 15 millivolts, focused attention brought the reading up to 3 volts; researcher Elmer Green found that healers produced voltages up to 190 volts! Gary Schwartz, professor at the University of Arizona, was intrigued by Green's experiments and began, what became a long line of, experiments finding that healers' greatest energy came from their dominant hand and produced a third more magnetic field changes than Reiki masters, proving that "directed intention manifests itself as both electrostatic and magnetic energy." Schwartz then became interested in the work of a German physicist, Fritz-Albert Popp, (with whom McTaggart later worked) who had proven that plants and humans "emitted a constant current of photons – tiny particles of light." He "theorized that this light must be like a master tuning fork setting off certain frequencies that would be followed by other molecules of the body."

Popp eventually formed the International Institute of Biophysics, composed of fifteen groups of scientists from around the world, including CERN in Switzerland and Moscow State University in Russia, which by early in the 21st century included at least 40 distinguished scientists. Popp and colleagues, building on leading edge research, found that biophotons measured from plants, animals and humans were highly coherent. (There's that powerful word again!) They acted like a single super powerful frequency, called 'superradiance' in some circles. "As one scientist put it, coherence is like comparing a 60-watt light bulb to the sun." If you could get all the photons in one bulb to become coherent and resonate in harmony, the energy would be thousands of times higher than that of the surface of the sun. Building on these discoveries, Schwartz realized that if thoughts are generated as frequencies, healing intention is well-ordered light. "Directed intention appears to manifest itself as both electrical and magnetic energy (humans are both receivers and transmitters) and produce(s) an ordered stream of photons measureable by sensitive equipment." "Like any other form of coherence in the subatomic world, one well-directed thought might be like a laser light . . ."[44] (from *The Intention Experiment*; McTaggart's latest book, *The Bond*, also published with Free Press, further illuminates her fascinating research)

Elisabeth Targ, Marilyn Schlitz and colleagues' 'Love Study,' as it came to be called, set up, logged and interpreted the results of an experiment that brought the remote healing intentions of a varied group of healers, from Christian to Native American shamans, to a group of AIDS patients. The results were amazing: the treatment group had six times fewer AIDS related illnesses and four times less hospitalizations than the control group, who was not sent healing intentions. McTaggart writes, "certain conditions and mental states make our intention especially powerful." Other experiments showed that all four lobes of the cerebral cortex took part in intuitive awareness, but the heart appeared to receive the information before the brain did. "People appear to receive healing deep in their bodies by being returned to the more coherent energy of the healer's intention."[45]

You have probably heard of the experiment that Herbert Benson, a cardiologist at Harvard Medical School, and team carried out with the

Tibetan Buddhist monks, who, in near freezing temperatures, were able to dry with their body heat, sheets drenched in cold water, three times! Monks and their meditation practices became the subject of in-depth studies questioning whether they had special neurological gifts "or did they acquire a skill that ordinary people could learn." McTaggart conducted questionnaires and interviews, finding that the first step was "achieving a state of concentrated focus, or *peak attention*." McTaggart further reports, "According to Stanley Krippner, an expert on shamanic and other native traditions, virtually all native cultures carry out remote healing during an altered state of consciousness and achieve (this) concentrated focus through a variety of means:" ayahuasca, drumming, chanting, meditation, prayer, even intense heat such as in a sweat lodge. "Eight of the Dalai Lama's most seasoned practitioners of Nyingmapa and Kagyupa meditation were flown to Richard Davidson's lab at the University of Wisconsin" and hooked up to 256 EEG sensors each to record electrical activity and asked to carry out compassionate meditation (the desire for all living things to be free of suffering). These were compared with a group of undergrads who had never practiced meditation before, who were likewise attached to the EEG. Surprisingly, according to the EEG readings, the monks' brains did not slow down, but began speeding up – on a scale no scientist had ever seen before. Sustained bursts of high *gamma-band* activity – 25 to 70 hertz, or cycles per second – indicated a state of rapt attention, deep levels of learning, ecstasy, and great flashes of insight, plus the brain waves began to operate in synchrony. (There is that 'coherence factor' again!) "The gamma state is even believed to cause changes in the brain's synapses, the junctions over which electrical impulses leap to send a message to a neuron, muscle or gland."[46] Different forms of meditation produce different brain waves. "It may be that the monks' compassionate intention, produced thoughts that sent the brain soaring;" those who had been performing meditation the longest recorded the highest levels of gamma activity.

Tests have proven that sustained thoughts (meditation, prayer) produced physical differences in the brain, and the healers all produced a coherence and synchronization between the two hemispheres as well as integration of the lower emotional center and the cortical seat of higher reasoning. Meditation classes are now held at many major corporations and have been shown to improve productivity and creativity. Some even have prayer groups and chaplains to help their employees. "A study at the University of Pavia in Italy and John Radcliffe Hospital in Oxford showed that saying the rosary had the same effect on the body as reciting a mantra . . . six times a minute."[47] This compares in one way to a recent book/CD put out by the Abraham-Hicks organization (which we will go into detail about later), *Getting Into the Vortex: Guided Meditations CD and User Guide*,[48] that paces one's breathing at about six breaths per minute. I have found it very helpful.

Functional magnetic resonance imaging (fMRI) measures the minuscule changes in the brain during critical functions, by calculating the increase in blood flow when certain neural networks are engaged. Herbert Benson and Sara Lazar, neuroscientist, used the fMRI, EEG and other monitoring equipment on people who had practiced kundalini meditation at least four years and found an increase in neural activity and the results suggested that highly concentrated focus might enlarge certain parts of the brain and "increases in cortical thickness were proportional to the overall amount of time spent meditating." When the gifted psychic, Ingo Swann, was examined, the MRI showed he had an unusually large parieto-occipital right-hemisphere lobe, the portion of the brain involved with sensory and visual input.

Other researchers found that exceptional healers visualize themselves uniting with the person to be healed, and then both being united with what they often describe as the absolute. (Supreme Being?) Others describe a sense of unconditional love, loss of identity or ego, or a mystical identification with the guardian spirits or guides. McTaggart found in her work with healers that some saw themselves as the 'water' or source of healing while others (by far the largest group) saw themselves as the 'hose' or channel for healing energy to flow through. Much of the imagery the healers used to describe what they did included relaxing, releasing, or allowing in spirit, light or love. McTaggart spoke with Targ about her work before she (Targ) died: "She found that a quality of loving compassion or kindness was essential in sending out a positive intention to heal." The healers said they would 'get out of the way,' surrender to a healing force and frame their intention as a request – *please may this person be healed.*[49]

Franz Halberg, known first as the father of chronobiology, now called chronoastrobiology, was the first to realize that the geomagnetic field as well as the sun and planets affect the rhythms of every biological process, including brain functions. We saw in the chapter "Astronomy" how disturbances in the geomagnetic field can increase seizures, attempted suicides and out-of-body experiences. "Although periodically destabilizing, exposure to the daily ebb and flow of earth's geomagnetic activity may be essential to life here."

Experiments in subjecting volunteers to low-frequency complex magnetic disturbances (similar to increased geomagnetic activity) via a 'helmet' blocking out all other electromagnetic noise, caused 'microseizures,' hallucinations, feeling of spiritual vision, epiphanies, and/or out-of-body sensations. Variations in the natural geomagnetic field or solar patterns appear to have an effect on extrasensory perception (esp) and the power of intention increases when the Earth's energy is agitated.[50] We saw earlier that some believe there will be a mass effect from the increasing electromagnetic energy coming into our solar system at this time.

Experiments with intentions directed toward small quartz crystal oscillators seemed to effect profound changes in the Zero Point Field, (explained previously) much like the 'conditioned space' found at sacred sites or the 'phantom healing charge' on any object, such as fabric, that the Hungarian healer, Oscar Estabany touched. All these "demonstrate the extent to which the energy of human thought can alter its environment. The ordering process of intention appears to carry on, perpetuating and possibly even intensifying its charge."[51] (Bear with me! We are working our way toward understanding why and how we may add our positive intentions to mass consciousness in the important time ahead.)

The power of the mind has been known for a long time in the 'placebo effect' and now is being utilized to perform surgery without anesthesia. "Placebo is a form of intention – an instance of intention trickery," says McTaggart. "Cases of spontaneous remission suggested that casual thoughts that run through our minds every day, together become our life's intention." "The most effective healing intention (was) framed as a request, combined with a highly specific visualization of the outcome." "These discoveries offer convincing evidence that all matter in the universe exists in a *web** of connection and constant influence . . ." "We must open our minds to the wisdom of many native traditions, which hold an intuitive understanding of intention. Virtually all these cultures describe a unified energy field not unlike the Zero Point Field, holding everything in the universe in its invisible web." McTaggart concludes, "The modern science of remote influence has finally offered proof of ancient intuitive beliefs about manifestation, healing and the power of thoughts." "Both modern science and ancient practices can teach us how to use our extraordinary power of intention."[52]

Wayne Dyer writes specifically about this subject in *The Power of Intention*. He asks, "*How do I go about getting what I intend to create?*" The answer: "You get what you intend to create by being in harmony with power of intention, which is responsible for all of creation." "You're looking for a vibrational match-up of your imagination and the Source of all Creation."[53] Dyer recounts the research of Dr. David Hawkins from *The Eye and I*, which tells of the impact of one person* and the importance of "matching up with energy of the universal Source:

- One individual who lives and vibrates to the energy of optimism and a willingness to be nonjudgmental of others will counterbalance the negativity of 90,000 individuals who calibrate at the lower weakening levels.

- One individual who lives and vibrates to the energy of pure love and reverence for all of life will counterbalance the negativity of 750,000 individuals who calibrate at. . .

- One individual who lives . . . to the energy of illumination, bliss, and infinite peace will counterbalance the negativity of 10 million people who . . . (approximately 22 such sages are alive today). (2004)

- One individual who . . . to the energy of grace, pure spirit beyond the body, in a world of nonduality or complete oneness will counterbalance the negativity of 70 million people who calibrate at the lower levels. (approximately 10 such sages are alive today)."[54]

Science and Spirituality

"All matter originates and exists only by virtue of a force...
We must assume behind this force the existence
of a conscious and intelligent Mind.
This Mind is the matrix of all matter."
- Max Planck, 1944

"With these words, Max Planck, the father of quantum theory, described a universal field of energy that connects everything in creation: *the Divine Matrix*." This is the opening of Gregg Braden's book with that title.[55] "The fact that this field exists in everything from the smallest particles . . . to (the) distant galaxies whose light is just now reaching our eyes, and in everything between, *changes* what we've believed about our role in creation." ". . . the existence of (this) primal web of energy that connects . . . everything in the universe opens the door to a powerful and mysterious possibility." "(A) possibility that suggests we may be much more than simply observers," one that hints that all our achievements, material abundance, loves, careers, fears, even beliefs are "made manifest through this mysterious essence of the Divine Matrix and that *consciousness itself* must play a key role in the existence of the universe." Before his death in 1992, David Bohm, Princeton physicist and colleague of Einstein, left theories "that opened the door to what he called the 'creative operation of underlying . . . levels of reality'." "It's from these subtler levels that our physical world originates," creating the universe as a single unified system of nature. Bohm's work was based on atoms in special, gaseous state called *plasma*, which, in this state, behaved as though they "were connected to one another as part of a greater existence."[56] He viewed both the seen (which he called the explicate order) and the unseen (the implicate order) as expressions of a greater, more universal order, a grand cosmic pattern, an undivided wholeness, or hologram. The deeper implicate order is the original and projects what we see manifested. In one experiment he witnessed free electrons organizing themselves in a plasma soup and described them as displaying mind-like properties. Gregory Sams writes along the same lines, "Just as all the life in the ocean depends upon the humble plankton for its survival, so might all complex expressions of consciousness in (the) Universe depend upon the simple intelligence of subatomic particles and light itself.[57] This helps us understand that Infinite Intelligence is part of everything from the smallest particle to the largest galaxy and, I would suggest, dimensions beyond what we know. Just as our DNA from any part of our body contains the pattern or code for the whole body, Braden explains, "The flow from the unseen (the Matrix, the 'soup of potential') to the seen is what makes up the dynamic current of creation."[58]

Gregg Braden compares this scientific view with a summary translation of *The Songs of the Sabbath Sacrifice*, from the Dead Sea Scroll fragments: "What happens on earth is but a pale reflection of that greater, ultimate reality." "As above, so below." It seems consciousness interacts with this unseen, omnipresent, everywhere 'realm' to create reality, then that creation shows us (mirrors) where we may have created beautifully or 'erroneously' and not to our satisfaction. Braden goes on to show that the oldest wisdom traditions (what we have called 'perennial truths' previously) knew how to communicate with this powerful realm he calls the Divine Matrix; it is through the language of human emotion. This Matrix has been called the Tao; "it is all that is – the container of all experience, as well as the experience itself." Our experiences of thought, feeling, emotion, belief, judgment, anger, feelings of separation are all disturbances in, and affect the Field or Matrix. Universal emotions such as love, hate, fear, doubt, and forgiveness are all part of the language we use to program the Matrix that creates our life; Braden encourages us to "hone our skills to understand how to bring joy, healing and peace to our lives."[59]

He emphasizes there is a field of energy that connects all of creation, it mirrors the beliefs within us, and we communicate with it through our emotions. "Ultimately, our survival as a species may be directly linked to our ability and willingness to share life-affirming practices that come from the a unified quantum worldview." He explains, our good wishes, thoughts and prayers are already at their destination (as we saw earlier – love travels faster than the speed of light), we are not limited by our bodies or the 'laws' of physics, we can support loved ones everywhere without ever leaving our home, we *do* have the potential to heal instantaneously, and it is possible to 'see' across time and space without ever opening our eyes.[60] And all that is in just the Introduction!

When we see and feel ourselves, body, mind and soul, so totally intermeshed in this seen and unseen, unifying energy 'soup' we will know the unlimited nature of The All-That-Is. Braden explains that "everything from our greatest fear to our deepest desire begins in this 'soup of potential'"[61] What we used to call 'miracles' will be understood as being part of what has always been possible, or as Jesus said, "He who has faith in me will do what I am doing, and he will do greater things still. . ." John 14:12 (The New English Bible, New Testament) Braden quotes Rumi in this regard: "What strange beings we are! That sitting in hell at the bottom of the dark, we're afraid of our own immortality."[62]

We are co-creators; we live in a world that is created by our consciousness. As the research at the Institute of HeartMath has shown, the heart center projects the vibration of emotion more than the head; a doughnut-shaped field of electromagnetic energy emanating from the heart extends five to eight feet in all directions. Braden further illuminates this research in *The Spontaneous Healing of Belief*, saying studies by this Institute has "shown that the electrical strength of the heart's signal, measured by an EKG, is up to 60 times as great as the electrical signal from the human brain measured by an EEG, while the heart's magnetic field is as much as 5,000 times stronger than that of the brain. What's important here is that either field has the power to change the energy of atoms, *and we create both* in our experience of belief!"[63] ". . . three attributes set the Divine Matrix apart: first, it is everywhere all the time; second, this field originated when creation did – with the big bang or . . . the beginning; and third, it has intelligence and responds to the power of human emotion." Braden reveals, ". . . ancient traditions shared this secret on temple walls and timeworn parchments; . . . the Gnostic Gospels used the word *mind* to describe this force – 'from the power of silence appeared a great power, the Mind of the Universe.' . . ." "It is the living essence that is the fabric of our reality, inside us and surrounding us."

Braden was able to question a wise Tibetan monk and was told the secret of what connects us to everything: "Compassion is what connects all things." "The Vedic traditions (also) speak of a unified field of 'pure consciousness' that permeates all of creation." "Scientists suspect that the relationship between mass prayer and . . . individuals in communities is due to a phenomenon known as the *field effect* of consciousness."[64]

This 'field' is so responsive to our every thought and feeling that our life constantly 'mirrors' those beliefs, words and attitudes back to us with what 'shows up' in our life. As Braden says, ". . . it is impossible to end war by creating more wars" and "angry people can't create a peaceful world." Master teachers such as Buddha, Jesus, Mohammed, Gandhi, Mother Teresa and Martin Luther King, Jr. used this principle by living their wisdom, compassion, trust, and love "within the very consciousness that they chose to change."[65] Albert Einstein said, "You cannot solve any problem in the same state of consciousness in which it was created." The teachers named above lived in the problems of one world but taught from a higher level of consciousness. The prince, Siddhartha, realized that his true nature was awareness (*budh* in his language), consciousness, or pure presence, and came to be known as the Buddha.[66]

Arjuna Ardagh also says, "most everyone agrees we live in a world gone slightly mad." He has interviewed hundreds of what he calls 'contemporary translucents' and describes them as those who "appear to glow from the inside. They have access to their deepest nature as peaceful, limitless, free, unchanging, and at the same time, fully involved in the events of their personal lives." He places Eckhart Tolle, Byron Katie, Ram Dass and Jean Houston in this category. "According to the most conservative estimates, millions of people today have been touched by such an awakening." His "2005 book, *The Translucent Revolution* . . . presents a portrait of what "may potentially be a new kind of human being." "However severe our challenges as a race, we are also in the early stages of a huge shift in collective consciousness, which is gaining momentum every day."

After this book came out, Ardagh and his wife were invited to visit the Oneness University in India and came away convinced that this simple 'blessing' in which the 'blessing-givers' placed "their hands on the recipient's head for less than two minutes," could be the miracle we were hoping for. Readings using EEG and SPECT scans point to reduced activity in the brain stem, sometimes called the reptilian brain, a balancing of hormonal secretions governed by the limbic cortex, a reduction of activity in the parietal lobes, an increase in activity in the frontal lobes and an increase in whole brain coherence, indicated by increased neurological activity in the corpus callosum." "What makes the Oneness Blessing remarkable," writes Ardagh, "is that these changes . . . happen in a matter of minutes rather than decades." Even a single blessing begins the process and with further Oneness Blessings, the changes continue to deepen and stabilize. Sri Amma and Sri Bhagavan see that as the Oneness Blessing spreads, and enough people 'Awaken into Oneness' there will be a shift in the collective consciousness of humanity from separateness to oneness. The year: 2012.[67] We now see science is proving what the spiritual leaders have been saying for centuries.

The Universal Law of Attraction

At last we come to what has been one of the most powerful and clear paths of guidance in consciousness I have ever experienced: that of the Teachings of Abraham™ (no, not the one in the Old Testament) channeled by Esther Hicks. *Abraham* is the name that Esther found closest to the vibration/thought form she received from this group of loving, non-physical entities, who variously refer to themselves as Source energy or spiritual guides. These guides explained that it was Jerry's (Hicks) desire to find answers to his many questions and Esther's pure openness and ability to quiet her mind, plus a pre-incarnation agreement, that led to this powerful relationship. In a session with *Theo*, channeled by Sheila Gillette, mentioned previously, Esther asked what she should do to meet her spiritual guide; she was told to meditate fifteen minutes a day, concentrating on her breathing. During meditation she first felt as though something very loving was 'breathing with her.'[68] From there to automatic typing to allowing *Abraham* to speak and teach through her was a short journey. Esther was really not familiar with channeling prior to meeting *Theo*, but soon appreciated the Infinite Intelligence, love and brilliance that *Abraham* brought through her. She became a 'bridge' between the physical and Non-physical worlds. They (*Abraham*) virtually removed that 'veil' between worlds as the Mayan shamans may have done in their 'journeys' between the Underworld and the Earth.

Abraham began by sharing comprehensive teachings about the eternal Universal Laws that encompass both the physical and non-physical worlds as well as their wisdom about our total beingness which also merges both worlds. Their explanation of our Non-physical part that they call the *Inner Being* (some call it our Higher Self or Soul) and our physical self, and the relationship between them, sheds much light on how our consciousness functions. Your Inner Being is not some mystical hazy, fairy-tale personage. It is you, the you that is never negative, never fearful, never full of doubt, never resentful, angry, sad, depressed, despairing or confused; it is the part of you that is always connected to Source Energy, Light, always loving, compassionate, joyful, content, clear, sure and knowing, just to name a few adjectives. In their words: "It is our desire that you return to an understanding of the immense value that you are to *All-That-Is*, for you are truly on the Leading Edge of thought, adding to the Universe with your every thought, word, and deed. You are not inferior Beings here trying to catch up, but instead, Leading-Edge creators with all of the resources of the Universe at your disposal."[69] Very powerful words!

"The *Law of Attraction* says: *That which is like unto itself, is drawn.*" (It is) "the most powerful *Law* in the Universe" and you must understand it before you can utilize the other universal laws. "It is the basis of everything that you see manifesting . . . the basis of everything that comes into your experience." "When you say, 'Birds of a feather flock together,' you are actually talking about the *Law of Attraction*." ". . . when you see that the one who speaks most about prosperity has prosperity . . . you will begin to recognize the exact correlation between what you have been thinking about and what is actually coming into your experience." "*You attract it – all of it. No exceptions.*" They explain, "Because the *Law of Attraction* is responding to the thoughts that you hold at all times, it is accurate to say that *you are creating your own reality.*" "Whether you are remembering . . . observing . . . or imagining, the thought that you are focused upon in the powerful now has activated a vibration within you – and the *Law of Attraction* is responding to it now." "*Without exception, that which you give thought to is that which you begin to invite into your experience.*" "By focusing on (an) unwanted thing, or the essence of it, you have created it *by default.*" "To clarify, when you look at something (or think about it) and shout, "No, I don't want . . . that; go away!" then what you are actually doing is calling it into your experience" (by your attention to it). "*You get the essence of what you are thinking about, whether it is something you want or something you do not want.*"[70]

Thoughts have magnetic, attracting power and draw other thoughts, experiences, conversations, and things of a similar nature to you. In the twelve or fourteen years that I have been listening to the Abraham-Hicks tapes, CDs, DVDs and reading their books, I have seen the teachings evolve as those listening have evolved. As *Abraham* says, "Our work is continually evolving through the questions that are being asked . . . There is no end to the evolution of that which we all are."[71]

Esther Hicks started sharing *Abraham*'s teachings with others in the mid-1980s and now does a full schedule of seminars in the U.S. and several other countries, including Australia, as well as week-long seminars on two or three cruises each year. She and her husband, Jerry, have published nine or ten books (I'm not sure I have all of them!) with Hay House, and have offered hundreds of tapes, CDs, DVDs, affirmation card sets, calendars, etc., all from the information channeled from the Non-physical group, *Abraham*. I attended a seminar in Albuquerque, NM several years ago and left so uplifted and inspired from seeing Esther translate the wonderful information she was receiving from *Abraham*. The Hicks have devoted their lives to this work and have quite a following. The questions that people ask are practical and the answers so user-friendly. I cannot do justice to the scope of their work in this short space, so I highly recommend their monthly CD program, their books and seminars, as well as their website: www.abraham-hicks.com

The quote from the King James version of the Christian Bible "Ask and ye shall receive," is so important to our understanding of the Law of Attraction. Gregg Braden expands it with the original Aramaic text that is translated: "So ask without hidden motive and be surrounded by your answer – Be enveloped by what you desire, that your gladness be full." "With these words, we're reminded of the quantum principle telling us that feeling is a language to direct and focus our consciousness," says Braden. "It is in the coupling of the imagination – the idea of something that could be – with an emotion that gives life to a possibility that it becomes a reality. Manifestation begins with the willingness to make room in our beliefs for something that supposedly doesn't exist . . . through the force of consciousness and awareness."[72]

In the Teachings of Abraham™ "the *Science of Deliberate Creation* says: 'That which I give thought to and that which I believe or expect – is.' In short, you get what you are thinking about, whether you want it or not." "Whether you are thinking of what you do want, or . . . thinking of the lack* of what you want (the direction of your thought is your choice), the *Law of Creation* goes to work upon whatever you are thinking about. "Your hospitals are filled to the brim with those who are now taking action to compensate for inappropriate thoughts. They did not create the illness on purpose, but they did create it –through thought and expectation." "By *default*, you are offering your thought. . ." We all know from our computer that unless we change the settings our computer program will automatically go to the *default*.

What is the default setting of feeling most of us get up with in the morning? "Oh, my back hurts. Already I'm running late! Why can't people drive right?!" It's not just the words, but the emotions (anger?) and feelings (fear?). If they're not lined up with those of your Inner Being, who is always in the positive, peaceful, loving zone, there will be a 'non-connect' that feels bad, throws you back into your negative default, and creates your life in ways that will not be as you wanted. We go around like little porcupines whose quills are all on the negative side of the 'magnet' and when something good or positive tries to get in our life, it is quickly repelled away by this negative force and all the 'bad' stuff we do not want in our life is pulled right toward us! We can 'override' the neural patterns or chemistry with a 'flood' of appreciation and reset the chemistry or pattern to a new 'default' of Wellness and Happiness. At least that is how I see it!

In the last few years, *Abraham*, through Esther, have centered their teachings around what they call our *Vortex of Creation*, where our Inner Being exists, and where our every 'rocket of desire' and positive intention goes, is 'vibrationally' created and held there until we align our thoughts and feelings with that and bring it into physical manifestation. Reminiscent of Carlos Castaneda's work, their emphasis on deliberate 'intent' helps us to know how important it is to hold and focus our positive thoughts and visualizations. Over the years *Abraham* has offered many processes that help with creating in your mind the 'things' or situations you want to have in your life. Emotion is the key. ". . . become specific enough in your intentions (and visualizations, I might add*) that you bring forth positive emotion, but not so specific that you bring forth negative emotion (feelings)."[73]

This relationship between you and 'YOU' – your Inner Being – your Source energy, is the only liaison you ever have to be concerned about. As long as you 'line up,' and are 'in agreement with' that (YOU), you will have perfect guidance. Just like a GPS made JUST for you. As *Abraham* tells us, we knew when we decided to come into this time/space reality that there would be 'contrast,' events, situations, and/or people that would not be to our liking, but we also knew that all we had to do was turn our attention toward what we wanted and our Inner Being would go to work virtually, creating that for us – immediately. All we have to do is stay 'honed in' on that 'signal.'[74] I am paraphrasing their work, of course; they go into countless details in their many books, CDs and DVDs, answering the weightiest as well as simplest questions from the entire gamut of life experience. Their book, *Ask and It Is Given* tells the story in its title, and says, "Well-Being" is lined up outside your door. Everything you have ever desired, whether spoken or unspoken, has been transmitted by you vibrationally . . . (and) has been heard and understood by Source and has been answered, and now you are going to *feel* your way into allowing yourself to receive it . . ."[75] So, now you can see the reason I first tried to show how 'everything is vibration,' and how consciousness (thought) is an energy and has an effect on those basic, quantum building blocks of our physical universe.

Gill Edwards, a clinical psychologist and spiritual teacher, also speaks along these same lines in her article called "Wild Love Sets Us Free" (in *The Mystery of 2012,* Sounds True) in which she says, ". . . love is the great healer; (it) overcomes the illusion of separation. When love is present, there is no fear. When there is no fear, everything is welcomed into your heart. . ." "Love reassures us that all is well. It sets us free. Love unites and connects, whereas fear disconnects." ". . . this time of great awakening (that is, the (time) leading up to and beyond 2012) is all about love. It is about expanding our consciousness . . . and realizing the cosmos is based upon unconditional love, from which we are inseparable." ". . . we are not victims* of fate, chance or karma, but divine cocreators of everything (without exception) that happens to us."

She goes on to say, "Religion has been used as an agent of social control and conformity." "Instead, we are divine and creative beings . . . we are sparks of God in conscious evolution." "At the cutting edge of physics and biology . . . the universe is an interconnected web* of energy-consciousness." ". . . we discover a universe in which love is so unconditional that it says,

'Yes!' to any desire we have. No questions asked. No need to earn or deserve it. . ." "The universe has infinite resources, and it can coordinate the higher needs of everyone. Ask and it is given." Edwards insists, "We get what we focus on. Focus on problems, and they get worse. Focus on your fears, and they will eventually manifest. Focus on what you desire, and you pull it toward you. . . What you expect, you get. You *can* have whatever you want, but you need to send out consistent signals. Your beliefs, desires, and expectations need to be coherent*."[76] (There's that lovely word again.)

Edwards quotes from the Abraham-Hicks® materials, ". . . the basis of life is freedom, the purpose of life is joy, and the result of life is growth."[77] ". . . a spirituality of unconditional love, in which each of us is a creative spark of an infinite and omnipresent Source of energy-consciousness . . . is why I urge people to choose to be happy instead of trying to be good," Edwards says, quoting from her book, *Wild Love*,[78] or as Joseph Campbell put it, 'Follow your bliss.'" "Gandhi said we need to *be* the change we want to see in the world. Because all energy is interconnected, we are all one at an energy level . . ." "Our energy radiates far beyond our personal lives, like ripples spreading out across a great lake. When we are negative or critical, even in the privacy of our own minds, that energy affects the world. When we are loving, joyful, peaceful, creative, and visionary, that, too, affects the world around us." She concludes this article by saying, "And, almost without noticing it, the boundaries between heaven and Earth have melted away."[79]

I knew the teachings by Abraham-Hicks seemed familiar, like meeting an old friend again, but didn't make the connection until, during research for this book I ran across the work of Arnold M. Patent. I had read *You Can Have It All* years ago. His most recent book, *The Journey*, sounds as though it would 'travel' along perfectly with us on this 'journey' we are taking. His website is: www.arnoldpatent.com

Perennial Wisdom

In the "Foreword" to *Healing States* by Alberto Villoldo and Stanley Krippner, Lynn Andrews (author of *Jaguar Woman* and several other books) states, "We each need our own unique healing . . . It is true that each of us, though not . . . a shaman or healer, is on a journey,* seeking wholeness . . . our own enlightenment." Andrews writes, "Perhaps this is the true reason we are on this mother earth . . . There is a healing sound inherent in each living creature in this universe. The challenge . . . is to create our own dynamic rhythm."[80] Sound and rhythm imply, to me, a 'vibration' or 'tone' of consciousness that is creative as we saw above. Villoldo and Krippner use the "word 'consciousness' to mean a person's overall pattern of perceiving, thinking and feeling. A 'state of consciousness' refers to the pattern that exists at any given point in time. "Some states . . . are said to be especially conducive to self-healing or to the healing of others."

They visited and researched the 'spiritual healing' techniques of Dr. Eliezer Mendes (he was a surgeon before establishing his clinic outside São Paulo, Brazil), who treats people with epilepsy, schizophrenia and multiple personality disorders, having learned his methods from the Umbanda religious tradition. In less than a month of work with hypnosis, past lives, the desire to be healed, as well as teaching the 'partner' of each patient to be a 'medium' for their loved one, most go home completely healed. ". . . the native healing traditions teach that health (is) maintained through healthful lifestyles, connection to the Great Spirit and the Mother Earth, and through service to others and to the planet." These authors emphasize, "We can take the more traditional paths . . . or we can take the path of 'planetary healing,' becoming positive and transforming influences in our schools, homes and workplaces and developing impeccability in our ethics and actions."[81] Again, we are reminded of Carlos Castaneda's work, from the word 'impeccability.' His early work with shamanism brought a new vocabulary to the field of consciousness.

"Psychology tells us . . . we must understand what traumas we suffered as children, or how our dysfunctional parents taught us unhealthy behaviors. But dissecting the past is a trap," Villoldo insists in *Illumination*. "The shaman knows that focusing on our wounds will only reinforce them . . ." "The shaman knows that whatever beliefs you hold about the nature of reality, the universe will prove you right." (Remember the *Law of Attraction?*) ". . . positive beliefs help you to see the glass as always half full."[82] "Opposites may attract in the world of (classic) physics, but in human relationships fear breeds terror, and generosity attracts plentitude." "Regardless of whether it is an inner or an outer journey* . . . you will be blessed and graced."

"For Buddhists, illumination is a state free of suffering and rebirth known as *bodhi*, literally 'awakening.'" In mythology and in life ". . . in a moment of euphoria, enlightenment . . . (an) awakening must be followed with the great departure and a journey* to a new destiny." Villoldo calls these levels of awakening *Initiation* in this book, *Illumination*, and points out that in this process "the pre-frontal cortex of the brain begins to regulate the primitive brain and its fear-based programs. The relaxation response will wash over your body . . ." "There are two great leaps in awareness that we make during our *initiation*: the awakening to our mortality, which happens when the neocortex first turns itself on; and the awakening to our *immortality*, which happens when the prefrontal cortex comes online."

"All deaths, all endings, are a passage to the next beginning. The first law of thermodynamics states that energy can be changed from one form to another, but it can neither be created nor destroyed." He goes on to say, "A Lakota medicine man once told me that when the ancestors die, they're buried and return to the earth to form part of the trees and all of nature, reassimilated into the greater whole. For a long time I thought this meant that their bodies became fertilizer, mulched into food for trees and flowers, but later understood that their consciousness endured and became part of the whole of Creation once again."

"Once we understand that all death is merely a transition into a different form of life, change ceases to be so frightening," says Villoldo. Thinking about life and death as a cycle is not new; many ancient cultures had symbols and ceremonies marking the connection between the old and the new, the dead and the living."[83] This was certainly seen with the Mayans, especially in the New Fire ceremony. "But if you see that you're an essential, indispensable part of Creation, you understand that you're responsible for creating your experience in this life. You'll stop believing that creation happens from the top down, with an all-wise supernatural being directing traffic on those rare occasions when 'he' feels like getting involved in the world," teaches Dr. Villoldo. "Instead, you'll understand that you're essential for the creative process and that Spirit is looking to you for input and direction. The next task for the shaman is to gather her personal power so that she can draw upon it to create."[84]

Non-duality and Consciousness

"The struggle between good and evil is the primal disease of the mind" - a quote from Seng-ts'an in Stephen Mitchell's beautiful book, *The Essence of Wisdom*.[85] "The end-date alignment can be thought of as an eclipse, and it shares with eclipses the basic alchemical meaning of 'the transcendence of opposites*,'" writes John Major Jenkins. "In Mayan metaphysics, this union has a more profound meaning that goes beyond the union of male and female and other pairs of opposites. Instead, it involves the nondual* relationship between infinity and finitude, eternity and time – a union of lower and higher." "Our higher and lower natures are reunited in eclipses, in the Quetzalcoatl myth of the sun joined with Venus, and the 2012 eclipse of the galactic center by the solstice sun. Our higher nature does not destroy our lower nature, but embraces it. The manifest world of appearances is restored to limitless possibilities when human consciousness reestablishes its connection to infinity." Jenkins concludes, "This is the heart of the promise of world renewal in 2012, and it can only take place within the heart of humanity." "These principles of Mayan sacred science are really perennial truths."[86] (from *The Mystery of 2012*)

The union of Father Sun with the Cosmic Mother should help us, symbolically, with our own tendency to put everything in opposing camps: order/disorder, passive/ aggressive, Islamic/Christian, black/white, Democrat/Republican, instead of bringing them together at a higher level. "The aboriginal mentality recognizes no ultimate distinction between matter and spirit, no dualistic split between mind and world." ". . . traditional cultures believe that human beings are inseparable from natural laws and cosmic cycles,"[87] writes Daniel Pinchbeck. The Abraham-Hicks materials teaches, "You are Source Energy in physical bodies and you are vibration at your core. You make a stark distinction between physical and non-physical. There is no distinction; there is a vibrational basis to all (both). At your core, the Source within you, who you refer to as non-physical, as well as the physical, is (all)* vibration."[88]

Perhaps we should look to the 'lower' level of subatomic particle/waves from which everything emerges. Gregg Braden's "Belief Code 22 *"Our belief in one force for everything that happens in the world, or two opposite and opposing forces – good and evil – plays out in our experience of life, health, relationships, and abundance,"*[89] is quite to the point! He writes, "Just the way sound creates visible waves as it travels through a droplet of water, our "belief waves" ripple through the quantum fabric of the universe to become our bodies and the healing, abundance, and peace – or disease, lack, and suffering – that we experience in life. And just (as) we can tune a sound to change its patterns, we can tune our beliefs to preserve or destroy all that we cherish, including life itself."[90] In quantum physics wave/particle is one.

"Polarization has a way of fixating our attention and pulling us into its fragmentation and division. The sheer scale of corporate control and military budgets can paralyze us with feelings of powerlessness . . . but like buds that have held tight in their own growth, humanity will blossom. Our blossoming will be the emergence of a consciousness no longer trapped in the old dualities," writes James O"Dea, president of the Institute of Noetic Sciences and former director of an office of Amnesty International.[91]

Imbalance in the left and right hemispheres of the brain has added to the duality in the world, or has the duality around us caused our imbalanced brain? This is where meditation comes in: to bring balance and coherence to our brain and consciousness. Chogyal Namkai Norbu, Dzogchen master, (born in eastern Tibet in 1938, former professor of Tibetan culture, Naples, Italy) teaches that overcoming dualism is essential to raising awareness, as it is the root of our suffering and our conflict. What is this dualism that we must put an end to? Encarta lists four definitions, two seem to apply here: first, a state in which something has two distinct aspects which are often opposites, and second, the religious doctrine that two opposed forces of good and evil determine the course of events. Many transformational authors are stressing that in order for humanity to thrive during this change from one world age into the next we must put aside polarizing beliefs and find common ground to agree upon.

Holism' recognizes that both the *creationist* notion of a pre-existing pattern and the *evolutionist* theory of how this pattern manifested over time, is an example of how opposing sides can come together to form a new image. We have had the tendency, especially in the West, toward competition vs. cooperation, science vs. religion, Right vs. Left, male vs. female. But we see in the yin-yang symbol, containing both back and white and entangled in such a way to show both are made of the same elements, that each contains the seed of the other. As Lipton and Bhaerman show in *Spontaneous Evolution*, we have examples from Buddhism that "love of self and love of others are one and the same. Compassion is an expression of human freedom, flowing from a sound intuition of the unity of life and all living things."[92]

Vimala Thakar in her beautiful little book, *The Eloquence of Living*, writes "We live in duality and know unity only as theory. We are all too familiar with the endless pressures of feeling apart, isolated, and the tiresome worries of being in competition with our fellow beings, but we don't know the marvelous ease of living, the soft rhythms, when the myth of separation dissolves." "Awareness melts the apparent divisions between living and dying, silence and sound, light and dark. One sees life as a whole, unfragmented, just as one sees the water of an ocean undivided though thousands of people have made thousands of designs on its surface. The indivisibility, the unfragmentable totality is the beauty of life."[93]

In *The Gospel of Thomas,* one of the Nag Hammadi texts found in upper Egypt, we find: "When you make the two one, and when you make the inside like the outside and the outside like the inside, and the above* like the below,* and when you make the male and the female one in the same . . . then will you enter the kingdom of God." The word 'humanity' should mean 'humans in unity.' Whitley Strieber, quoting The Master, wrote of Jesus, known as the Christ, "his imperfect humanity and his perfect divinity were the same thing." "I am here to get you to recognize yourself as divine. . ." "The new man will live in ecstasy, even though he lives in chains." "God is all being." "God is you."[94]

Another modern spiritual teacher who encourages a direct experience of the truth is Mooji, as he is known to his devotees. Born Anthony Paul

Moo-Young in Jamaica; he moved to the UK in his teens. His meeting with a Christian mystic led him to a direct experience of the Divine within and he was able to allow this to deepen and flower. He felt he had met an enlightened master in 1993 when he met Papaji (Sri Harilal Poonja) and became his disciple. Mooji has shared *satsang* all over the world since 1999.[95] He is said to assert that the two paths to truth of devotion and knowledge, can be brought together in union.

The Master (in *The Key*) taught that "the teachings of Buddha, Christ, and Muhammad are interlinked. They are one system in three . . ." "Christianity seeks God, Islam surrenders to God, Buddhism finds God. When you see these as three separate systems, you miss the great teaching of which each contains but a part. Seek the kingdom (within) as a Christian, give yourself to God as a Muslim, find your new companion in the dynamic silence of Buddhist meditation."[96]

A Common Vision

"**G**rowing numbers of thoughtful people are coming to the conclusion that *intentionality* directly shapes reality." "Effectively transitioning to this new world will require envisioning it into reality," recommends author John L. Peterson, founder of The Arlington Institute, a future-oriented think tank. "We will all need to develop a basic but coherent* idea of what the new world might look like . . . and begin to carry that common picture in our minds."[97] Llewellyn Vaughan-Lee states, "The image the first astronauts gave us of (our) world seen from space, a single beautiful sphere without borders or divisions as well as . . . the number eight (∞) as an important symbol of two worlds coming together, are two (images) . . . (that) carry primal energy . . . and have the power to . . . reconfigure the collective consciousness."[98]

Dr. Christine Page reminds us, "Your contribution, along with that of other awakened souls, will create a blueprint, which will influence . . . the next 26,000 years of human existence." "For the first time in 26,000 years, our sun . . . (is) aligned with the dark rift of our galaxy, the Milky Way . . . (which) represents, according to the ancient people, the great mother and the great serpent, both synonymous with death and rebirth. . ." "It is seen as the portal to multidimensional reality, parallel worlds, the void . . . the ocean of possibilities" ". . . *the anxiety of the unknown,* (is) often translated as 'chaos'." Page continues, "The prophecies say we are entering an era governed by the element of ether . . . often described as the synthesis of the four material elements – earth, air, fire, and water. Ether is celestial . . . existing beyond the confines of time and space . . . a fusion of polarities* that eliminates the need for separation into darkness and light, negative and positive, or good and bad. All are accepted as another expression of the same essence, lovingly created in the presence of the great mother." Page goes on to explain, "Within the medium of ether where time and space are collapsed into *the now*, all our actions are simultaneously met with a . . . response so that we automatically experience the effect of our deeds upon any person they impact. In essence, we will no longer be able to rest in the belief that our thoughts and emotions are safely hidden as long as they continue to be unexpressed and unspoken." ". . . we will come to truly appreciate the phrase, 'Do unto others as you would do unto yourself' for we will understand that there is no differentiation between the self and others."[99] Men and women alike will need to let go of their old stories, reach into their psyches with love to let go of shame and fear and fully embrace those parts.

Nan Moss writes, "Mystical, esoteric traditions understand the individual human's power to influence manifestations in this world. Eckhart Tolle, a contemporary mystic and philosopher, teaches that we are all part of one divine organism. As such, when we focus our attention on anything . . . we complete a circuit of awareness. Tolle asserts that Nature is waiting for us to wake up to this, and the beings of Nature are calling us to draw closer to them through our hearts and attention." (*Weather Shamanism*)

"Ancestral traditions were well aware of this relationship of humans and the natural world and took great responsibility for their co-creative role. In the eyes of indigenous peoples, we are not really human until we understand this," she further explains. "We all have the ability to experience different states of consciousness; it's part of our internal structure." "In the shamanic worldview, all that exists is alive, and everything and everyone is interrelated with everything else." This is the *Web* or *Tapestry* that we have spoken of before. "Shamanism," Moss explains, "recognizes that we live in a universe of both *ordinary reality*, this physical world of space and time, and *nonordinary reality*, the usually hidden spirit world. Knowledge of both these worlds is vitally connected to the well-being of all." She suggests that modern-day shamans probably have more of a struggle than shamans of the indigenous cultures.

"We can all learn how . . . to help restore balance and bring to light how to conduct our lives in sustainable and harmonious ways . . . and begin (to) expand . . . our worldview into a more inclusive paradigm."[100]

She relates that shamanic practitioners (as Moss is) must understand that the need and desire for harmony between our world and the world of Spirit is mutual, and I believe this is important for all of us to comprehend. Moss closes her book, *Weather Shamanism*, with a poem by Ramona Lapidas, called "Spider Woman Speaks," which was received during a shamanic journey, in part: *"You have all that you need within you to add a greater light to the web. / Go / Dream me a new dream, / Weave me a new web,"*[101]

Barbara Marx Hubbard spoke in 2005 of the "tragedies that are occurring in the lives of so many people in so many ways at the same time that many new possibilities are being born." She called them 'concurrent realities.' "They create the terrible dichotomies within which so many people now live. The only way that anyone could hold a positive vision of the future would be to see our present-day circumstances in spiritual terms." Neale Donald Walsch related this in his 'uniography', as he called it, of this well-known figure in the social transformation movement, *The Mother of Invention*.[102] We shall hear more from both of them in our Conclusion.

Our Energy Fields

Alberto Villoldo's mentor/teacher/shaman, Don Jicaram, taught, ". . . memory is not kept in the brain – neither is consciousness, for that matter – but in the body and in the fields of energy that surround the physical body." In the training he did with this shaman, Villoldo learned of the various 'bodies' one acquires in learning the 'levels' of consciousness that are attained when working with the Four Winds of the Medicine Wheel: the energy body, the Nature or etheric body, the astral or "mystical body (Wisdom of the universe)" and the causal body – the thought before the action, the creative principle.[103] He also spoke of the 'luminous energy field' or information field which surrounds the body from his work with don Ramon and other shamans of the Amazon. This part of a person's energy field could be cleansed of 'heavy' or toxic energies or 'issues' such as fear, past life traumas, or illness with their ritualistic actions.[104]

"In Mayan culture, the aura is sometimes symbolized by feathers – a representation of the realized being." In Mayan the word *czin* (or *cizin)* refers to the crown chakra, and means radiating energy. *Cizin* is also their word for enlightenment, auric energy or illumination. "To become Quetzalcoatl is to know how to consciously use this radiation with all our awakened faculties." "A flowering of radiation is visually represented by the serpent symbol of solar energy."[105] In *The Key* The Master (as author Whitley Strieber refers to him) describes an "electromagnetic field that fills the nervous system (and) rests a few centimeters above the skin . . ." "It may be imprinted by information from anywhere and any time."[106] Let's look at some other ways of understanding these energy fields that surround each of us.

The part of Carlos Castaneda's work that left the most lasting impression on me was his teachings about the energy field around the body that his mentor, *Don Juan Matus,* called our 'luminous egg.' Whether you consider Castaneda's body of work as an addition to anthropological discovery or a great fiction, it did bring an entirely new frame of reference to the consciousness movement. Self-development writer, Wayne Dyer, wrote that Castaneda's last book helped him realize that the 'power of intent' was not just a self-help technique, but a force that exists as an invisible field of energy. He begins Chapter One of *The Power of Intention* with a quote by Castaneda, "In the universe there is an immeasurable, indescribable force which shamans call intent and absolutely everything that exists in the entire cosmos is attached to intent by a connecting link." Dyer goes on to say, "In my mind, intention is something much greater than a determined ego or individual will." "Imagine that intention is not something *you do*, but rather a force* that exists in the universe as an invisible field of energy!" "I know that intention is a force that we all have within us." Dyer goes on, "The Source, which is intention, is pure, unbounded energy vibrating so fast that it defies measurement and observation. So, at our Source, we are formless energy, and in that formless vibrating spiritual field of energy, intention resides." "When you lay ego aside and return to that from which you originally emanated, you'll begin to immediately see the power of intention working with, for and through you. . ." "All you have to do to tap in to the power of intention is to be in a perfect match with the Source. . ."[107]

Chris Griscom, visionary, spiritual teacher, past-life therapist, and founder of The Light Institute in Galisteo, NM, speaks of the 'emotional body' – an energy field around us that carries the memory of our past lives. Although I had accepted reincarnation for several years, I had a chance to feel its reality when the medical world said there was nothing more that could be done for my husband's lung cancer and we started looking for some kind of alternative cure. When none was found, we turned to the healing of the soul/body. I had read several of Ms. Griscom's books and remembered that past life regression therapy had helped many people with various life issues. Though Paul's strength was fading, we were able to schedule four days of regression with a wonderful facilitator at the Light Institute. He helped us, separately, examine five of our past lives, two in which we had been together before. In four of the lives Paul examined he had been in a violent relationship with the person he had known as his father in this lifetime. In viewing these in his 'mind's eye' he was able to let go of the anger he had felt for so long but had not fully understood. His soul was healed even if his body was too far gone. I cannot say that seeing my 'self' in other lives/situations, all of which included dying in various ways, was 'life-changing' but I did perceive death in a different way and was able to go through the final two months with my husband in a less fearful way.

I recommend Chris Griscom's early books (I have not read her latest books) and The Light Institute (at least, as I saw it several years ago). She writes, "We, as a species, are making a pivotal leap of consciousness into oneness, a consciousness in which we dissolve the illusion of separation from the Divine."[108] Griscom's *Ecstasy is a New Frequency*,[109] though written in 1987, is still a valuable resource for those seeking to know their inner and outer selves better and pull the two together. Her work reminds me of the work of Abraham-Hicks that we discussed above. We must come to know ourselves as extensions of Source energy or what we have called the 'Divine.' The *Vortex* spoken of by *Abraham* may be another field of energy – that of our Higher Self, or 'spirit body.' Just a thought! Or perhaps a good visualization.

Reincarnation

A Cherokee elder offered this information: "We knew that, because of the disaster that would befall this planet, there would be those who would be needed to call upon their reincarnation memories to heal, to counsel and to love a world gone mad, a world without knowledge, a world without hope. . ." ". . .it will be the duty of all those who seek spiritual knowledge to instruct others when the Earth moves. . ." "It was determined that, through these (crystal skulls), the minds of oneness would be activated."[110] So, although reincarnation may seem strange or new to those familiar with only the Judeo-Christian viewpoint, it is the basis of belief for millions (perhaps billions) of people worldwide, especially Buddhists and people believing in Hinduism. It seems important to our 'journey' into the subject of consciousness to discuss the eternal and evolving nature of the soul.

Daniel Pinchbeck reports, "Ian Stevenson, former professor of psychiatry, University of Virginia, spent forty years studying young children who spontaneously recall(ed) past lives, compiling data from 2,600 cases in his book, *Reincarnation and Biology*. He found numerous incidences in which children recalled specific details of their last life, and has documented . . . birthmarks and birth defects that seem to carry over from their former existence . . . especially those in which the past life ended in violent death, the birthmark perfectly matched the placement of the mortal wound. Most of the children spoke of the previous life with . . . strong emotion. Stevenson documented past life memories among the Tlingit in Alaska, the Druze in Southern Lebanon, across India and other regions." Pinchbeck quotes Stevenson, "The recollection usually commences between the ages of two and four and ends within the next five years." He corroborated specific memories with family members of the previous life. "The average period between death and rebirth (was) fifteen months (in this study)."[111] (*2012: The Return of Quetzalcoatl*)

Pinchbeck also quotes Rudolf Steiner who "said the specific mission of his life on Earth was to bring the knowledge of reincarnation back to the West." "In their remote mountain kingdom, Tibetan Buddhists developed, over centuries, a highly evolved spiritual science of reincarnation." "The current Dalai Lama was identified, as a young child, as the reincarnation of the previous one, through oracles and foretellings. Steiner believed that a belief in reincarnation has been continuous over the course of human development. His esoteric philosophy was thoroughly evolutionary, proposing that everything in the cosmos perpetually transforms – not only human beings and planets, (as well as) the higher 'spiritual beings' . . . and even the basic laws of the cosmos . . ." "He saw the universe as a staging ground for infinite transformations and permutations of consciousness. . ."[112] I particularly like what Eckhart Tolle has to say about reincarnation: "Reincarnation doesn't help you if in your next incarnation you still don't know who you are."[113] So true.

The Mayans believed that Quetzalcoatl (or, perhaps, his spirit) would return in the 'end times' to guide humans through the transition into a new era. Carlos Barrios, Mayan teacher, says, "Many great sages have been reincarnated in different parts of the world, and they will be the ones to bring comprehension and harmony. The times we are living in require great clarity. These times require a tradition that is pragmatic, clear and congruent with Mother Nature." "Freedom is living in harmony, where every human being respects himself or herself, others and Mother (Earth)." "The Fifth Sun is a time of harmony, peace, tolerance, and balance." "Humanity will have access to a subtle dimensional form, and cosmic law will prevail . . . the *B'alameb'* will return for this period . . . (for) these beings . . . taught this humanity and are the cosmic guardians of the four corners of the universe." [114]

"There is no death; it is only the relinquishing of one form for another. As energy is constant, only the form changes," states *THEO*, channeled by Sheila Gillette, ". . . in many parts of the world, most believe in the ongoing soul essence." When asked about 2012, they (THEO) replied, ". . . the year 2012 is significant. Not that there will be a dramatic upheaval. It is a pinnacle point of change, of awarenesses." "Human's diversity . . .

is a beautiful tapestry* (on) your Earth." "But it is important not to hold mind in thought of time. If (one is) fully present moment to moment in one's experience, one lives the life to its fullest extent."[115]

"The early Greeks, who received much of their knowledge from Egypt, believed that souls reside in the Milky Way between incarnations, and there are two 'gates' . . . the Silver Gate of Gemini, through which souls descend to Earth; and the Golden Gate of Sagittarius, through which souls ascend,"[116] writes Geoff Stray. In *Hamlet's Mill* we find, "The ancient Mangaians of the Austral Islands of Polynesia regarded the Milky Way as "the road of souls as they pass to the spirit world." In their myth, souls are not permitted to stay unless they have reached a stage of unstained perfection, which is not likely to occur frequently. "Polynesian souls have to return into bodies again, sooner or later."[117] I first came to really believe in an afterlife when I 'saw' (in my mind's eye) my father-in-law come into our bedroom briefly – two hours after he died 500 miles away, to check on his son. I have never doubted since then that we are eternal beings and have an 'afterlife.'

Chris Morton and Ceri Louise Thomas relate that their native guide at the ruins of Tikal, in Guatemala, explained that the "Mayans and other ancient tribes of Central America had a different understanding of death from our own. To them death was not a full stop . . . not something to be afraid of, but rather something to look forward to, a great opportunity to pass into another dimension, a chance to join with the world of the spirits and the ancestors. To these ancient people death was part of the balance of nature, part of giving back to Mother Earth the life she had given. The skull (carved into the side of one of the pyramids) was symbolic of this view.[118]

This brings up the related subject of karma, which Sylvia Browne explains is a 'sister' of dharma. "The word 'dharma' originated in the ancient language of Sanskrit . . . and at its most basic, it means 'protection.'"[119] "Protecting ourselves from unhappiness is no one else's responsibility but ours, and it's accomplished not by physical force but by living within our own inner laws of righteousness, peace and tolerance. Dharma dictates absolute honor for all living things and the land that nourishes them. . ." Browne stresses, "When we live outside the bounds of our own divine dharma, we're naturally out of synch with the universe and pay the price in the form of stress, misery and bitterness. . ." "Karma is another universal law that boils down to sayings (like) "What goes around comes around," and "What ye sow, so shall ye reap". . ."[120] Motive, intent and refusing to accept responsibility for one's actions come into play here. I have heard recently that as consciousness evolves and time seems to be speeding up, our karma comes back to us much faster. Although I have written a novel (set in the 16th century) around the theme of karma and reincarnation, I doubt I fully understand all the nuances of either. I do believe that what one puts out in thought, words or actions attracts back to him/her those same essences in this time/space reality, if not in one life, then in another and we all knew this before we were born. You know this in your 'soul'. It is another part of the Universal Law of Attraction. Eckhart Tolle writes, "Most people are at the mercy of that voice (in the head); they are possessed by thought, by the mind. And since the mind is conditioned by the past, you are then forced to reenact the past again and again. The Eastern term for this is karma."[121] Perhaps the soul carries this subtle 'thought form' about unresolved issues or negative feelings and, if not resolved, they can cause physical or mental manifestations.

"Both Mayan and Christian sources talk about an end to death at the end of time, and Eastern traditions talk about the enlightened state as deathless. The ancient traditions all point in the same direction: a timeless, enlightened state of cosmic consciousness." "The Mayan calendar will have come to an end, and its use becomes meaningless." "We will be completely free to chart our own destiny. Humanity will live in true freedom, joy and peace," predicts Dr. Carl Johan Calleman.[122] Perhaps the word 'death' will disappear from our vocabulary as we will understand that we are eternal beings transitioning in and out of physical (dense) manifestation. In the Abraham-Hicks materials *Abraham*, through Esther, likes to use the word 'croak' for the death experience since there is no death, as they say. "We are eternal beings."[123]

Whitley Strieber relays that the Master of the Key said, "Energetic bodies hunger to be radiant. They taste of ecstasy and want desperately to find their way to the completion of joy . . . (they) need to return to time to reconstruct what of themselves impedes their ecstasy." ". . . the mission of this age . . . is to open the elemental (physical) body to ecstasy." "This is the age of God within."[124] Regardless of our view about many incarnations, we are quite sure we will have only one chance in this personality, the one I know as 'Lataine,' in this space/time 'reality show' to grow, learn and evolve – to do what we came here to do this time around.

The Power of Thought

Are we beginning to see ourselves in a broader format, as powerful, eternal beings, connected to the physical as well as the non-physical worlds, as multidimensional? "Psychoanalyst Carl Jung realized . . . the interconnection of mind and world, the psychic and the physical, expressed through synchronicities (another word I like very much) and other occult correspondences." Jung understood that a deep transformation of the human psyche was underway, writes Daniel Pinchbeck. "According to Jungian theorist, Edward Edinger, we are currently experiencing the "archetype of the Apocalypse," a term that has the familiar destructive connotations but is also defined as 'a revealing' or 'an uncovering.' As a negative archetype, it represents the smashing* of previous forms of thought and ways of being; as a positive archetype, it represents a momentous event – the coming of the Self into conscious realization,"[125]

Pinchbeck goes on to assure, "If it is the case that mind and matter are increasingly interpenetrating as we approach the 2013 phase-shift, then our level of consciousness and the power of our intention* will determine how events play out for us, and upon the greater world stage." "We best serve the transformational process . . . by attaining a level of being that is without judgment, anxiety, or negative projections."[126]

Dr. Jean Houston sees the current stage of collective growth as being propelled by five forces: (1) We are waking up to the realization that there is an evolutionary pulse "by Earth and Universe toward a new stage of growth." "As ancient peoples have always known, the story is bigger than all of us . . ." (2) ". . . capacities that belonged to the few must become the province and requirement of the many* if we are to survive the next hundred years." (We must) "aspire . . . to a much large awareness." "The consciousness that solves a problem cannot be the same consciousness that created it." (I believe Einstein said something similar.) (3) "The movement seems to be from the egocentric and the ethnocentric to the worldcentric – a fundamental change in the nature of civilization. It raises hope for forgiveness between and healing among nations and ethnic groups . . . an empathy . . . that honors the golden rule of human interchange." "We need models of a new order of relationships . . . one in which male and female, science and spirituality, economics and ecology, civic participation and personal growth come together in an . . . interdependent matrix for the benefit of all." (4) Barriers dissolve as more people are coming to accept the benefits of cultural diversity. (5) "Not since the days of Plato, Buddha and Confucius has there been such an uprising of spiritual yearning . . . more and more people are gaining access to the source of our being and becoming." Six billion, nine hundred nine million (Census Bureau, March 30, 2011) "members of the human family congregated together on a spinning ball, in stress, in ferment . . . prompts us to articulate goals lofty enough to lift us out of petty preoccupations and unite us in pursuit of objectives worthy of our best efforts." "The world is hungry for vision." From her chapter in *The Mystery of 2012* (Sounds True) Houston concludes, "It is a matter of *kairos* . . . ancient Greek for that moment when the shuttle passes through the openings in the warp and woof threads," a potent time when the new fabric can take form.[127] This is a beautiful way to describe the tapestry and web that we collectively weave with our lives. Seven billion people make a very large 'entity' of mass consciousness and a huge force of creation.

George A. Seielstad, author of *At the Heart of the Web*, quotes Loren Eiseley (famous anthropologist and naturalist, author of *The Immense Journey*): "Man lies at the heart of a web, a web extending through the starry reaches of sidereal space, as well as backward into the dark realm of prehistory . . . It is a web no creature of earth has ever spun before. . ." "Even now one can see him reaching forward into time with new machines . . . until elements of the shadowy future also compose part of the invisible web he fingers." Beautifully, poignantly said. Seielstad extrapolates, "If the web's time dimension does extend into the future, it will be because humans learned to regard life on earth as a *totality*, a single superorganism integrating several complex ecosystems into a whole that is greater than the sum of its parts." ". . . we humans now have the capability to guide and shape evolution into a cosmic future. So profound is this change that few are aware it has occurred. Each of us may be analogous to a single cell in a human body; just as that cell is not aware of the consciousness possessed by the total system to which it contributes, an individual human may not be aware of the cosmic consciousness possessed by the totality of his species. Nevertheless, those of us alive at this moment of unique cosmic significance must sense its epochal magnitude . . ." "The evidence that all earth's biota are woven into a common web* manifests itself in the way genetic information is stored . . ." "The storage of genetic information is the same whether the organism is single- or multi-celled, whether it is plant or animal, ancient or modern." "The genes in all these cases are immensely long molecules called nucleic acids. DNA, deoxyribonucleic acid, is the . . . substance that determines the particular form in which a living organism appears. Every cell of that organism contains closely similar DNA molecules." "The strength of life's ability to adapt, hence survive, has always been in its diversity." [128] More to let us know that we are 'all the same,' as well as unique.

While we are on the subject of DNA, Gregg Braden writes, "It appears that the vibratory template of emotion actually 'touches' the molecules of DNA in our cells, 'waking up' dormant codes of immunity and vitality that may lie dormant within us." "The way we resolve our emotion, programs our DNA, affirming or denying life within our cells in response to our beliefs and attitudes."[129] All the more reason to 'get happy'! See the glass, not as half-full, but **FULL**, perhaps overflowing!

"We are what we think. All that we are arises with our thoughts. With our thoughts, we make the world." The Buddha said that and Dr. Alberto Villoldo opens his book *Courageous Dreaming* with that thought. "Like the Australian Aborigines, the Earthkeepers (shamans/wisdom keepers of the Americas) . . . know that all of creation arises from, and returns to, this creative matrix. The dreamtime infuses all matter and energy, connecting every creature, every rock, every star, and every ray of light or bit of cosmic dust. The power to dream, then, is the power to participate in creation itself." He continues, "The Earthkeepers . . . in the Andes and Amazon believe that we can only access the power of this force by raising our level of consciousness." "... when we experience our connection to infinity ... we're able to dream (create) powerfully."[130]

Patricia Mercier explains that shamans "know that their awareness interacts with the environment in which they live." "By staying in close touch with the mystery of the Earth, indigenous peoples have an unspoken communication with their surroundings ... (knowing) that at some level we are all interconnected. In every cell of their bodies they honour Great Spirit." "Sometimes the Maya call this solar consciousness, or *cosmovision*, or attaining the qualities of the *Ahau,* the Sun Lord. In superconsciousness, truths of the greater mystery hidden within life, the Unknown, become instantaneously available." (In) "superconsciousness – all is One."[131] In *The Maya End Times,* Ms. Mercier reports, "The Incas of Peru like the Aztecs – each vanishing civilization - left keys to their knowledge. They knew that, at some time in the future, others would ... use these keys to open doors to higher knowledge, to the ocean of superconsciousness." "... like the ancient Egyptians ... they had hope, even a certainty, that those coming after them would evolve so that they would use the keys to bring forth an even more beautiful creation in the cosmos... a new consciousness."[132]

Pierre Teillard de Chardin (French Jesuit priest and paleontologist, 1881 – 1955) served as a stretcher bearer in Morocco in WWI, trained at the Sorbonne and received a science doctorate in 1922. Though required by The Church to give up lecturing at the Catholic Institute, sign a statement withdrawing his controversial doctrine on original sin, and continue his geology research in China, Chardin continued his writing, though none was ever accepted by the Vatican. *The Phenomenon of Man*, published posthumously, gives his vision of the Omega Point in the future, which is "pulling" all creation towards it, today called convergent evolution. He wrote, "In Omega we have ... the principle we needed to explain both the persistent march of things toward greater consciousness, and the paradoxical solidity of what is most fragile." "Self-reflection (consciousness) is, literally the turning in of a flux of love-energy upon itself."[133]

Geoff Stray notes in *2012 in Your Pocket* that Chardin concluded from his study of paleontology "that the purpose of evolution was to build up a mind-layer around the Earth" which Chardin called the *Noosphere.* "The process would culminate in an "Omega Point" (which) "would awaken and become a 'Christosphere.'" Stray worked with Sergey Smelyakov, a

Ukrainian mathematics professor, and Jan Wicherink on the final *Auric Time Scale*, which seems to carry Chardin's theory even further, and found that the decreasing spiral of the accelerating time map would reach an implosion point on December 21, 2012 – a virtual 'white hole.'[134] Does this mean time itself would be changed or just perceived differently?

John Major Jenkins also speaks of Chardin's Omega Point, explaining that it "suggested that humankind was approaching a new unification of consciousness on a higher level. . ." He goes on to say in *The 2012 Story*, that this is much like Terence McKenna's Singularity predicted to birth on 12/21/12. Jenkins explains that infinity is being progressively revealed and that this is very different from Darwinism in that it means, like all Perennial Philosophy, this is a tops-down concept – all change descends from above. "The implications of the Omega Point, the Singularity, and 2012ology require a radical revisioning of Western philosophy's approach to reality." ". . . consciousness doesn't evolve, it remembers, or awakens, to its full potential . . . physical brains evolve in order to accommodate awakened minds." He states later in this book that he thinks, "the Maya believed that galactic alignments* are involved in a potential awakening experienced by human consciousness."[135]

Sages through the Ages

God and the Evolving Universe states, "In India's great scripture the Bhagavad Gita, the warrior Arjuna becomes a powerful instrument of God by transcending his selfish needs and his attachment to immediate results. In Taoist teaching, it is said that by concentrating vital energies, quieting thoughts, and disregarding external rewards, we can realize mastery in everyday work and thus express our deeper nature..."[136] Mihaly Csikszentmihalyi, born in Croatia in 1934, who emigrated from Hungary to the U.S at age 22, has been described as the world's leading researcher on positive psychology and wrote the book, *Flow*. James Redfield and co-authors relay what Csikszentmihalyi says about 'flow': that it includes "deep concentration, clarity of goals, loss of sense of time, lack of self-consciousness, and transcendence of the sense of self." "Czikszentmihalyi's findings resonate with the claims of sages through the ages..." "The winds of grace are always blowing," said the Indian mystic Sri Ramakrishna, "but to catch them we have to raise our sails."[137]

Another body of ancient wisdom is that of Patanjali who may be, like Hermes, a mythical character; the name means "a gift from heaven that falls into the open palms of the yogi." Some scholars say he lived in second century BCE, others say the 5th century AD. Dr. Alberto Villoldo in his book *Yoga, Power, and Spirit*, which is his version of Patanjali's work, pulls East and West together when he says, "Just as shamans do, the Yoga Sutras advocate a knowledge that is experiential and personal. The word *yoga* derives from the Sanskrit word *yuj*, meaning 'to yoke or join'; and refers to our awakening to our nature as Spirit..." "The word *sutra* literally means 'thread'; and while Patanjali provides us with the wool, we must spin it and weave it into our own tapestry*..." (There's that beautiful symbol again!) "In the Bhagavad-Gita, written 500 years before the Yoga Sutras were composed, Krishna reveals the eternal yoga to Arjuna, explain(ing) that this yoga is ancient, and that he'd revealed it earlier to the sages of old, who conveyed it to the wise *rishi* kings." One of my favorite sutras in Villoldo's version is number 32:

> "Practice samadhi
> and the dance of action and reaction,
> inspiration and discouragement,
> suffering and joy,
> will end,
> for they will have served their purpose."[138]

This appears to advocate an end to duality and polarization that we looked at previously. Dr. Wayne Dyer also works with *How to Know God: The Yoga Aphorisms of Patanjali* in his book, *There's A Spiritual Solution to Every Problem*. "Patanjali teaches that we are capable of reaching a state of awareness in which we can perform miracles. He explains that we are transcendent beings to begin with and counsels us to be unafraid of transcending the limitations imposed upon us by the material world." "Patanjali offered hundreds of specific suggestions and practices to reach the oneness or union with God, which he called yoga." In his Introduction to this book, Dr. Dyer gives ten points for solving problems; I am going to quote five, which seem to reiterate themes we have seen previously in this book. "First: Everything in our universe is nothing more than energy. That is, at the very core of its being, everything is vibrating to a certain frequency. Second: Slower frequencies appear more solid and this is where problems show up. Third: Faster frequencies such as light and thought are less visible. Fourth: The fastest frequencies are what I am calling *spirit*. Fifth: When the highest/fastest frequencies of spirit are brought to the presence of lower/slower frequencies, they nullify and dissipate those things we call problems."[139]

Toward 2012 (Pinchbeck and Jordan) also quotes Patanjali: "The connection to the earth should be steady and joyful. Our relationship with all beings and things should be mutually beneficial if we ourselves desire happiness and liberation from suffering." Stephen Duncombe writes that following this "change(s) our approach to life from asking, "How can the earth benefit us?" to "How can we benefit the earth?"[140]

Working with shamans in the Amazon, Daniel Pinchbeck intuitively received a lengthy discourse from a 'spirit' that said it had incarnated as Quetzalcoatl in the Mayan period, that he/she occasionally takes human rebirth to accomplish a specific mission. "As foretold, I am also the Tzaddik – 'the righteous one' and the gatherer of the 'sparks' of the Qabalah . . . a form or vibrational level of consciousness." "Soon there will be a great change in your world." In summary Pinchbeck writes, "What is this universe? It is a poem that writes itself. It is a song that sings itself into being. This universe has no origin and no end. What you are currently experiencing is . . . a transition between two forms of consciousness, and two planetary states." "The first principle of my being is unconditional love." "Hell is a state of mind. When you eliminate fear and attachment, when you self-liberate, you attain the Golden Age." "What manifests outward from the ground of being is freedom in time, and freedom from time." "Thought generates new potentials and possibilities of manifestation. Thought changes the nature of reality. Thought changes the nature of time."[141]

Let's look next at what we have learned about the quantum nature of the universe, the many phenomena that are 'blurring' the line between worlds, and the wisdom of the Mayan shamans and other indigenous people that can be used in our daily life if we understand the laws of the Universe.

Global Coherence and Mass Consciousness

We've observed how powerful thought is - especially positive, coherent thought patterns. When that is multiplied by thousands and millions of people, it begins to have a strong global effect. One group that is working toward that positive global effect is the **Global Coherence Initiative** through their research, monitoring, and by providing a global meeting place on their website: www.glcoherence.org. As they state: "The Global Coherence Initiative is a science-based, co-creative project to unite people in heart-focused care and intention, to facilitate the shift in global consciousness from instability and discord to balance, cooperation and enduring peace." The Institute of HearthMath, founded by Doc Childre, is the sponsor of GCI. There is encouragement to subscribers to their website to send clear, focused, compassionate thoughts to various areas of the planet that are under extreme stress due to war, natural catastrophes or anything that brings people to crises.

In *Spontaneous Evolution* the authors write about GCI: "The power of group coherence was illustrated on May 20, 2007, a date designated as Global Peace Meditation and Prayer Day. More than one million people in 65 countries meditated and prayed for peace at a synchronized moment in time. The results were similar to those random number generators that shifted into coherence during events like Princess Diana's funeral or the attack on the World Trade Center. Global Coherence Initiative researcher Roger Nelson reported that monitors around the world recorded measurable increases in the coherence RNGs (random number generators) while the meditation was in process. Yes, coherent consciousness impacts

Earth's fields!" "Could the vision of a coherent civilization whose collective consciousness is focused on love, health, harmony and happiness truly create a field strong enough to manifest Heaven on Earth?"[142] As I've said before, "Sounds good to me!" In their Global Care Room there is a format to join with others three times a day (or any time) to send caring thoughts and prayers on the same subject. Their articles on the effects on humans of geomagnetic activity, sun spots and electromagnetic radiation alone have been worth the effort of the free registration. GCI recently sponsored a teleseminar with Barbara Marx Hubbard about her vision for launching a web connecting engine on December 22, 2012 that 'will help birth the new humanity.'

I will only give a few thoughts and quotes from Neale Donald Walsch's *Mother of Invention*, (2011) which is about the above-mentioned visionary, because I want you to read it! He calls this book a 'uniography' as in "the life story of an individual that turns out to be the life story of the Unified Whole, with a prediction of things to come arising out of things that have passed. . ." "Barbara Marx Hubbard believes that those who connect with each other and collaborate are helping . . . to give birth to that more co-creative, universal humanity. . ." Walsch has written nine books of dialogue of his *Conversations with God* so he "understand(s) if Barbara feels that she has communicated with Someone Outside Herself – who also resides *within*." She said to Walsch, "God said to me that God is talking with all of us, with every single human being, all the time."[143]

Because of my upbringing I am not quite as comfortable as these two people are with the word 'God,' having had to overcome in my early adulthood the image of the patriarchal, avenging, 'old man up there' type of god. And we will talk more about the names of God later so we can all become more comfortable with our relationship to the Creator or as the Mayans said, *Hunab K'u*, 'The One Giver of Movement and Measure.' Barbara Marx Hubbard seemed to be guided toward her role as "a storyteller of the planetary birth experience." As she told Walsch, "I've come here to act not as an exemplar, but as a catalyst – for so many others who are right at the threshold of their own emergence." "Because it is time . . . we have to learn ethical, conscious evolution quickly . . . (learn) to love each other, and nature, as ourselves."[144]

Besides visionary, catalyst, potentialist, and storyteller, Ms. Hubbard was a candidate for the office of Vice-President in 1983 to run with Walter Mondale. Geraldine Ferraro eventually was chosen as the Democratic running mate, but Barbara's platform was visionary, as usual, based on *holism* – the whole person, the whole community, the whole world. "A new process *of* the Whole, *by* the Whole, *for* the Whole," she declared. "We are at the beginning of the age of the whole . . . becoming an interdependent world." She proposed a *Peace Room*, under the direction of the Vice-President with equal status to the War Room.[145] In light of what has happened since that date, this would have been a great idea to initiate.

"Our consciousness is the *sum total* of what is contained in the Heart, Mind, and Soul. Because it is, truly, *everything* – every awareness, thought, idea, and belief ever held by every sentient being (and by the Source of Sentience itself) – we can access consciousness at many levels." "It is *attention* that keeps us glued to a particular place in consciousness." "The trick, then, is to get out of your head and into your heart," teaches Walsch. Hubbard's early influences were Betty Friedan, Abraham Maslow, Pierre Teillard de Chardin, and Jonas Salk – the last three she knew personally. Maslow helped create the human-potential movement; Chardin, as Barbara says, "saw God in evolution as the expression of a Divine process." "He said that humans are about to awaken together . . . in consciousness and in capacities," she reported, and "a jump in our own consciousness and freedom is possible and, indeed, natural." She liked that Chardin called this new kind of human that was evolving – 'Homo progressivus'."

Barbara believes in this collective Birthing of a New Humanity "by which humanity will emerge as . . . a magnificent form of our species . . . a form that only loves, and never again hates, that only shares, and never again hoards, that only heals and never again hurts, that only births and rebirths itself . . . and never again kills." To that end she is working toward unveiling on the Internet a site called the *Synergy Engine* that will bring together all of the world's best ideas and solutions for global coherence and connection. Launch date? December 22, 2012![146]

Already on the Internet is the *Global Consciousness Project (GCP)* which "is an international effort involving researchers from several institutions and countries, designed to explore whether the construct of interconnected consciousness can be scientifically validated through objective measurement.

The project builds on excellent experiments conducted over the past 35 years at a number of laboratories."[147] The 'live dots' all over the world showing coherence or non-coherence in that area are fascinating ongoing research. On their new items list I see that on March 31st Masaru Emoto (author of *The Hidden Messages in Water* and *The Shape of Love*) invited the world to join in a focused Prayer for the Water of Japan at noon each day. GCP has monitored a number of other mass meditations. Another item reported the death of Sathya Sai Baba, the guru, on April 24, with as many as 500,000 people expected to attend his funeral in India. It is said his devotees number in the millions. The global coherence graph around the time of his transition seemed to show a negative trend at that time. Under the heading "A Planetary Smile" is an article by Barbara Marx Hubbard and speaks of her development of *Birth 2012* and her work with Stephen Dinan on *The Shift Network* with a link to these on the GCP site. I particularly like the GCP link to their Poetic History section, where I found this entry:

WE ARE ONE PEOPLE
WE SHARE ONE PLANET
WE HAVE ONE COMMON DREAM
WE WANT TO LIVE IN PEACE
WE CHOOSE TO PROTECT AND HEAL THE EARTH
WE CHERISH THE EARTH'S BIO AND CULTURAL DIVERSITY
WE WILL CHOOSE TO CREATE A BETTER WORLD FOR ALL
WE WILL BECOME STEWARDS FOR THE PLANET'S THREATENED
AND ENDANGERED SPECIES
WE WILL DEFEND AND PROTECT THE RAIN FORESTS, REDWOODS, AND OTHER SACRED PLACES
WE WILL DO OUR BEST TO MAKE THIS DREAM COME TRUE
WE WILL DO THIS FOR OUR CHILDREN, AND OUR CHILDREN'S CHILDREN
WE WILL TRANSFORM WHAT NEEDS TO BE TRANSFORMED

WE WILL BREAK FREE OF OUR CHRYSALIS LIMITATIONS
WE WILL JOYFULLY LOVE, SHARE, AND FORGIVE
SO THAT PEACE MAY PREVAIL ON EARTH!
MAY PEACE PREVAIL ON EARTH!

Our Earth Proclamation, contributed by Alan D. Moore[148]

Jonathan Goldman's *Healing Sounds* website is another that is using the connectivity of the Internet to focus global coherence. Their "World Sound Healing Day" on Valentine's Day was reported by GCP and the graph of their monitors indicated a 'peak' in coherence shortly after the noontime event. Goldman is an international authority on sound healing, author of several books and a Grammy nominee. His book *The Divine Name* with its instructional CD, is about much more than intoning the Tetragrammaton (Yahweh) from the Hebrew Kabalistic practice, although that is very powerful. He gives instruction for 'toning' the chakras and tells of the power of the sound of vowels, considered sacred in many spiritual practices.[149] It may appear that chaos reigns in the world today, but there are signs everywhere of people of many persuasions coming together to make a more peaceful world.

Well-known Mayan researcher, John Major Jenkins, believes the Mayans were well aware of the 'rebirth' of humanity at the end of the 'Fourth Sun'. He writes, the "shaman-astronomers believed that specific types of alignments in the cycle of precession stimulate evolution for life on Earth." "The Maya envisioned the alignment to occur in 2012 as a union of the Cosmic Mother (the Milky Way) with First Father (the December solstice sun). Woven into Maya astronomy, mythology and cosmology is a profound understanding of Earth's evolving consciousness." Father Sun's movement into union with Cosmic Mother's heart (26,000 years ago) signified the insemination and, true to the field effects of the galaxy, we have been in 'spiritual gestation' which will reverse in 2012 and the birth of something new will occur.[150]

Other perspectives of this momentous event come from Geoff Stray: one is Ken Carey's 'moment of Quantum Awakening' which is described as "an out-of-body experience into a timeless realm caused by a magnetically induced pineal-eye opening . . . an evolutionary quantum leap that may be seen as the boundary between *Homo sapiens* and our descendant, *Homo spiritus*." Archaeologist Laurette Séjourné has interpreted the end of the Fifth Sun of the Aztecs in a similar way: ". . . there are various indications that the Fifth Sun (Fourth Sun of the Maya) is the creator of a great and indestructible work: that of freeing creation from duality." "The essence of the Nahuatl religion is contained in the revelation of the secret which enables mortals to escape destruction . . . by becoming converted into luminous bodies." Whitley Strieber calls this 'the radiant body.'[151]

Stray also reports on the 'geopsyche' concept, introduced by Persinger and Lafreniére's *Space-Time Transients and Unusual Events*, which proposes that at a certain critical number of biological units, a matrix is formed which takes on certain patterns and properties of its own. Visualize our areas of high population density in this theory; then, "assuming the 'energizer' to be consistent with previous geomagnetic storms or electrical disturbances . . . the phenomenon would be anywhere between a few seconds to a week," say these authors. Stray goes on to postulate that "if an unusually powerful energizer causes a permanent change by triggering mass electromagnetic hypersensitivity, or by turning on the N-methylating enzyme, then we have a model for the mechanism of transformation." He also reports that other authors believe the Hall of Records in Cairo will be opened on December 21, 2012 causing a mass illumination of mankind.[152]

"Now, we mortals live in the Middle World of *Homo sapiens*; the Upper World is the domain of *Homo luminous* – the realm of Spirit. This is the celestial terrain of angels and archangels, and of enlightened ones, who are free from time and death. The Upper World is where you attain your divine nature, yet it's also where you discover the beautiful agreements that you made with Spirit before you were born," writes Alberto Villoldo. "Legends tell us that journeying* to the Upper World is the hero's journey." "The myths teach us that you can only reach the peaks of the Upper World in a healed state, free of the demands of the ego and filled with grace and integrity." "Mayan prophets foretold of a new humanity being born from the ranks of the Upper World in the year 2012. This evolution of human beings will involve you, as you'll be part of a quantum leap into becoming a new human species that will grow new bodies that will age, heal and die differently." Villoldo reports that the Laika shamans of the Incan tradition believe "All creation is benevolent, and it only becomes predatory when we're out of *ayni*." (balance or proper relationship) "The origin of *ayni* is in the mythology of the indigenous Americans, which says that the universe is benign and will order itself to conspire on our behalf when we're in proper relationship with it."[153] In his book, *The Four Insights*, Villoldo writes, "According to the prophecies of the Maya, the Hopi, and the Inka, we're at a turning point in human history. The Maya identified the year 2012 as the culmination of a period of great turmoil and upheaval, one in which a new species of human will give birth to itself. We're going to take a quantum leap into what we are becoming, moving from *Homo sapiens* to *Homo luminous* – that is, beings with the ability to perceive the vibration and light that make up the physical world at a much higher level – evolve not *between* generations but *within* a generation."[154]

James Redfield's books (*The Celestine Prophecy, The Tenth Insight, The Secret of Shambala,* and recently, *The Twelfth Insight*) are all focused toward raising global consciousness. He obviously believes that human culture will change, evolve, and discover more about the energy dynamics of the universe – "an ongoing spiritual transformation," as he says.[155] Drunvalo Melchizedek writes of the imbalance now within the Unity Consciousness Grid. "We must really take responsibility for our thoughts, feelings, and emotions now. For each one of us is a Dreamer. And what we dream (imagine, 'see') will become real in this world. This is the Mayan belief: that as we move closer to December 21, 2012, and February 19, 2013, (See chapter on Atlantis) the power of the Dreamer becomes stronger and stronger."[156]

Barbara Hand Clow who gave us much of our information about the catastrophes that brought the Mayans and other indigenous peoples to their 'knowing' about World Ages, is also very much in agreement with what we have discovered, saying, "Consciousness creates the world." She believes those catastrophes affected mass consciousness in a very negative way – some religions call it the *Fall*. "Thousands of years would pass before we could wake up, while Earth bided her time. Earth knew a potent cosmic infusion of energy from the stars would cause us to begin to vibrate with Nature again, and we are." Clow is an International Maya Elder with the Mayan Council of Chichén Itzá. About these invigorating cosmic waves that are reaching Earth now she says, "Whether individuals can receive the waves, and whether their subtle glands can activate from these powers, depends on whether they are focused or distracted; that is, whether they are grounded in their bodies." "Simply by tuning ourselves to the frequencies we all are capable of holding, we invite the divine intelligences into the human world." She later says, "I hardly need to mention that millions of meditators attain these levels of awareness day after day."[157]

With his background in geology, computer science, his knowledge of quantum physics and indigenous and ancient beliefs, as well as his travels worldwide to meet wise monks, shamans and elders, Gregg Braden stands in a rare place to help us pull our information about consciousness and perennial wisdom together. I guess that's why I like his work so much. He writes, "We embody the collective power to choose which future we experience." ". . . an ancient science stat(es) that we may change the outcome of our future through the choices that we make in each moment of the present." "Through our choice of peace in our lives, we ensure the survival of our species and the future of the only home we know." He relates that the 'ills' of society that we see around us cause some to believe in the doom of 'end times.' "Witnessing the same diseases, military conflicts, and extremes of nature, and referencing the same prophecies, those who subscribe to (another) viewpoint sense that a rare birth is occurring, an integral element of which is an equally rare shift within humankind." He states, ". . . this view suggests that we are entering a time of joy, peace, and unprecedented cooperation among the peoples and nations of the world."[158]

Braden related a wonderful metaphor from one of his Native American friends who revealed that over an entire valley the sage plants are all joined together through their root system – they are all one family. "As with any family, the experience of one is shared to some degree by all others." He draws the parallel, "Though we see many bodies that we believe are strangers . . . there is a single thread of awareness that binds us as a family – through a system we do not see." "Still, the connection exists as what some have called a 'universal mind': the mystery of our consciousness. Like the sage plants, we are all related during our journey* through this world. In consciousness, there is only one of us here." "Recent studies into the effect of mass prayer . . . document . . . (that) the quality of life for an entire neighborhood has been shown to be affected by the focused prayer of a few individuals."¹⁵⁹ The Transcendental Meditation organization has proven this over and over in the past 30 years.

Barbara Marx Hubbard wrote a chapter, "A Vision for Humanity," in the Sounds True compilation, *The Mystery of 2012*, in which she notes that crises precede transformation. "Problems are evolutionary drivers." She asks, "Is it possible that a new species has been gestating in the womb of self-consciousness through the great avatars, mystics, and visionaries for thousands of years, and that now, due to the crisis, this new human is being born in millions of us? It has been predicted as *Homo progressivus* (Teilhard de Chardin), *Gnostic human* (Sri Aurobindo), *Homo sapiens sapiens sapiens* (Peter Russell), *Homo noeticus* (John White), *Homo universalis* (my favorite). It is possible that the basic shift point in this prophecy is the emergence of the 'universal human' . . ." ". . . God is reproducing godlings," Hubbard asserts. "We are to become co-creators with the process of creation until we have joined together with others doing the same in a universe that is infinite . . ."¹⁶⁰ We saw earlier in this chapter that Dr. Alberto Villoldo used the term *Homo luminous* for what we are becoming as a species. In the Foreword of *The Maya End Times*, Chief Sonne Reyna, Yaqui-Carrizo-Coahuilteka Nation, entreats, ". . . enjoy Patricia Mercier's global spiritual adventure. And celebrate the cosmic birthing of *Homo Spiritus*."¹⁶¹

CONCLUSIONS

The Journey

*L*et's see where our 'journey' toward 2013 has taken us. We may never know for certain whether the amazing culture of the Mayans was informed and/or guided by an earlier advanced civilization or if the shamans drew their wisdom from 'vision journeys' into Infinite Intelligence. We do know that they, like other indigenous peoples, were very sensitive to and connected to the Earth, her cycles and the cycles of the cosmos. Out of this came their amazing calendars and their foresight to preserve this knowledge for those coming at the end of this world age. There are indications that the Mayas and Incas felt there would be a great change in humanity at the end of this Fourth Sun and we assume that means a change in consciousness.

We have correlated more recent discoveries about the physical universe with what we know about the Mayans, and our appreciation of their perception has increased. As we looked at other ancient and indigenous knowledge and predictions we were drawn to examine other dimensions, 'strange' phenomena, and other ways of looking at the interface and interaction between the physical and non-physical worlds.

All these led us to examine consciousness and how it affects and interacts with all these worlds. Over and over we are reminded of the power of thought. We were also reminded by quantum physics as well as the ancestors: "As above, so below, as within, so without." Now we can bring it all together – the above and the below – into All One with no duality. Your journey is my journey and the journey of everyone; we are all on this trip around the galaxy together.

Sandra Anne Taylor writes, "In order to understand why your journey has taken a certain direction, it's helpful to know what your soul's motivation may be. There are five main reasons why spirit chooses to come here. Knowing what has driven you to this earthly adventure can help immensely in your personal evolution and self-empowerment." She lists and explains: "1. Your Spirit longs to express itself. 2. Your Spirit longs to experience. 3. Your Spirit longs to learn and grow. 4. Your Spirit wants you to awaken to your sacred identity. 5. Your Spirit longs to love and serve." (*Truth, Triumph, and Transformation*)[1] There is another beautiful Creation Story in this book. When we think of our 'journey' we wonder which Path we should follow. I believe there are as many "perfect paths' as there are people. Each person's connection and journey to their Higher Self/Source is unique and ideal for them and no one else. Gregg Braden seems to say something similar, "Ultimately, you will grow beyond the teachings of the outer masters. The more of yourself that you know, the fewer teachings there will be. The fallacy of the* 'Path' is that there really is no path at all; there is only your experience."[2]

What I have been referring to as my Inner Being or Higher Self, Mike Dooley calls the 'Greater Self.' In *Infinite Possibilities* he uses the word 'journey' in a wonderful way, as though coming into physical/time/space is an adventure, almost like an exciting vacation from which we will eventually go 'Home.' He says, "The *Conscious I* does indeed create my reality *without* any interference from my Greater Self, *because this freedom and ability alone are what my Greater Self desired*: that my unique personality and perspectives would sojourn into time and space to savor and enjoy the magnificence of it all as led by my blossoming consciousness, desires, and preferences that emerged during the journey,* however they emerged."[3]

The Pleiadians, through Barbara Marciniak, say, "We speak to you as evolving humans – as ambassadors of light. You are, at this moment in time, poised on unprecedented discovery. You are ripe, through your own exposure and your own seeking, to make a choice. When we journey* into your reality, we learn as we watch the processes through which you resist, learn, and create life. You can consider yourselves, at this moment in time, at a summation point of life's journey." "When you understand that these roads take you to the great shifts in consciousness you are now accumulating and experiencing as a mass consciousness, it will catapult an ending to reality as you currently know it." "As a species of life, you are poised at that moment when you are required to make a leap of faith."[4]

Rev. Deborah L. Johnson, founder of Inner Light Ministries and The Motivational Institute, and civil rights activist, received by automatic writing, her *Letters from the Infinite, volume I: The Sacred Yes*, over a period of years. It opens with: "This book is a journey. It is a journey of the soul. It has been said that the longest journey is from the head to the heart and back to the head again." ". . .it is also meant to be a guide for the journey. The journey is experienced on multiple levels. On whatever level one finds oneself experiencing the journey, it is fine and appropriate for that person at that point in time. There is no need to compare . . . your journey with that of another. Your journey is authentic and unique." "These letters are tools for living."[5]

When I think of a journey I sometimes travel in my mind to a wide ledge that lies about a thousand feet below the peak of Mt. Wilson, which is 14,246 feet elevation, in southwest Colorado (my favorite mountain). The snow there can be several feet deep in winter and in spring two beautiful waterfalls come off the ledge and meet at the bottom of a 'V'. Water makes its journey there by way of clouds coming off the Pacific Ocean and across California or the Baja. The 'lift' causes them to drop some of their 'load.' I hiked to this ledge once in July and found patches of snow there and on the peak. This is where Coal Creek starts its journey – from the rain and melting snow on Mt. Wilson. Down thousands of feet to the West Dolores River, to the Dolores River, to the Colorado River, through Lake Powell and Glen Canyon Dam, and down to the Sea of Cortez, and home to the ocean. As Rainer Maria Rilke wrote, "There is only one journey. Going inside yourself." So let's examine this beautiful tapestry we are all weaving together with our life journeys.

The Web of Life

Do not think of the tapestry we are all weaving with our lives as two or three dimensional, but multi-dimensional including time now, time past, time future, and timeless. Each of us has a grid reference point that holds us in a time-space position, explains Patricia Mercier. The Maya shamans helped weave the time-threads that are the sum of all past experiences in this life and others. She continues, "Then there are the Threads of Time Future – bundles of possibility and potential destiny, or 'callings', each with its own particular 'colour', vibration or frequency." "By adopting a simple code of living we can take full responsibility for our thoughts and actions . . . and making these new connections . . . gives us a kind of personal 'grid reference point'." A woman elder reiterated to her Chief Seattle's famous words, "We do not weave the Web of Life. We are but a strand in the Web of Life. What we do to the Web we do to ourselves. All things are connected. Every part of this Earth is sacred to my people. . ." Another elderly Native American woman told her, "Nothing is separate in the Sacred Hoop of life. We are all connected within it, for what we do to one has an effect on the many." Mercier remains convinced that we can change even the direst predictions on the Threads of Time, with concentrated intention carried out by as many awakened people as possible.[6]

If you somehow accidentally set your default to hopelessness or powerlessness, no problem; you've done nothing 'wrong,' nothing that can't be fixed anyway. Well-Being is the most pervasive 'color' in our tapestry; just reset your default to line up with That. Put more appreciation, laughter and beauty in your life! Choose your thoughts carefully; or as Ms. Mercier says, "In the deepest part of your soul and spirit, develop a burning desire to understand how you (can) spin yourself a different thread of life in the (months) leading up to . . . 21 December 2012." "We can choose to focus (as the shamans did) on things that are good, those bringing joy and happiness."[7] The choice is ours!

Gregg Braden calls this web the Divine Matrix, "It is all that is – the container of all experience, as well as the experience itself. The Tao is described as perfect, 'like vast space where nothing is lacking, and nothing is in excess'." "The key is not only to understand how it works; we also need a language to communicate our desires that's recognizable to this ancient web of energy . . . (and that) is the language of human emotion." "It may be helpful to think of the Divine Matrix as a cosmic blanket that begins and ends in the realm of the unknown and spans everything between. This covering is many layers deep, is everywhere all the time . . . all that we know exists and takes place within its fibers."[8]

He goes on to say the discoveries of the 'new physics' further emphasizes the connectedness of everything and asserts that connection is because of us. The key, Braden says, to awakening this power is to make "a small shift in perception" – "we must see ourselves as *part of* the world rather than separate from it." "Within this cosmic exchange, our feelings, emotions, prayers, and beliefs at each moment represent our speaking to the universe. And everything from the vitality of our bodies to the peace in our world is the universe answering back."[9] Like a mirror to show you what has been communicated.

Neville, the 20th century visionary from Barbados, "suggested that it's impossible for anything to happen outside the container of consciousness." "If there is in fact a single field of energy that connects everything . . . then there can be no *them* and *us*, only *we*." "Feeling is the language that "speaks" to the Divine Matrix. Feel as though your goal is accomplished and your prayer is already answered." ". . . if we're conscious, by definition, we're creating," Braden concludes.[10] "Ask and Be surrounded by your answer." Jesus of Nazareth

"Each day, you're weaving the tapestry of your earthly experience, so it's far more productive to make your *entire life* your intention. Every part of your experience is worth the effort; and when you come from a grander point of view, all of the factors will seem to fall into place more easily. From this open and interested perspective you'll become aware of things you never even thought of – and you'll be empowered by the laws of attraction and all the other elements of creation,"[11] writes Sandra Anne Taylor quoted previously.

Another insight about our 'tapestry' comes from Gregg Braden: "Within Biblical references as well as other ancient texts, the term light refers to the *full spectrum of electromagnetic information*, including and not limited to, visible light; *all wavelengths* of radiant energy permeating all of creation throughout all universes. Light is energy. Light is information. Light is the fabric* of creation."[12]

The Key (Whitley Strieber) speaks on this subject: "You are weaving a tapestry of living memories. This is what the body of man is – a great weave of shimmering living cloth. It is full of all the hopes, failures, fears and attainments of the ages." "All lives are all completely present in this work of art."[13]

The Names of Source

Even when we throw up our hands in frustration – we are silently invoking a Higher Power, appealing for help from 'above.' There is just that tendency to reach upward, to look upward. The Mayans referred to Father Sky and the Creator. The name that we give to this Supreme Being may tell us a lot about our beliefs around that entity or force. YHWH is the ancient Hebrew name for God (the Tetragrammaton) with the modern reconstruction or transliteration of that being YAHWEH or Jehovah, which some feel is too sacred to be spoken aloud. The Old Testament, Exodus 3:13-14 states, "And Moses said unto God, Behold, when I come unto the children of Israel, and shall say unto them, The God of your fathers hath sent me unto you; And they shall say to me, What *is* his name? What shall I say unto them?" "And God said unto Moses, ***I Am That I Am***: And He said, Thus shalt thou say unto the children of Israel, ***I Am*** hath sent me unto you." (King James Version)

I read an editorial once that said, "Each time we use the words, *I Am*, we are invoking the God within ourself." When we use the words negatively as in, "I am angry" or "I am sick," it is like a prayer to the Creator, and we continue to create that situation within us. We search for God outside ourselves. Most of us don't realize that we are within God and God is within each of us. Nothing can be *outside of God*. Someone sent the following quote to me in an e-mail: WE ARE NOT HUMAN BEINGS GOING THROUGH A TEMPORARY SPIRITUAL EXPERIENCE. WE ARE SPIRITUAL BEINGS GOING THROUGH A TEMPORARY HUMAN EXPERIENCE. Penney Peirce in *Frequency: the Power of Personal Vibration,* quoted Wayne Dyer who wrote something similar, "Begin to see yourself as a soul with a body rather than a body with a soul."[14] ". . . we're incapable of anything other than a spiritual life. . . as beings of spirit, we're capable only of spiritual experiences." ". . . the activities of every day can't be separate *from* our spiritual evolution – they *are* our spiritual evolution!" This from Gregg Braden.[15]

As I wrote earlier, I have had to release the image of the vengeful God spoken of in the Old Testament that had formed during my childhood. Then I had to, for peace in my soul, let go of a belief in a wise old man God 'up there somewhere' who chose to save only those who attended certain churches and obeyed certain rules laid down by a certain translation and interpretation of certain writings that the Nicene Council had chosen in 325 AD. I had to try to find a way to relate to The Creator and come to know this Force of Love that was/is/and always will be inside and outside me. These writers have helped me do that.

Gregg Braden described asking for wisdom about the forces of light and dark, when on a trek in Peru this way: "I had affirmed many times before in prayer my belief in the single source of creation, the fundamental vibration, the seed tone of the standing wave that allows the hologram of life's patterns." A voice asked, "Do you believe I am the source of all that you know and all that is your experience?" ". . . my body responded with a mental *yes*. The voice echoed its reply, "If you believe in me, and you believe that I am the source of all that is, then how can you believe, at the same time, that anything you experience is other than me?" "The conditioned belief that darkness is a force unto itself, a fundamental power, in opposition to and separate from all that is good, no longer held any truth." He continues, "Fear, anger, rage, hate, jealousy, depression, control issues and violation are each expressions of darkness playing out in our modern lives." "I was able to see darkness as a portion of the whole, a part of the source of all that is . . ."[16] Because we are part of Source, through free will we may create or experience light or darkness; we are still within All-That-Is.

Braden continues, "Through our perceived* polarity of darkness and light, we have the opportunity to view ourselves from a different perspective . . ." ". . . darkness and light are not two distinct and separate forces at odds with one another. Rather, each represents a portion of *precisely the same whole*, the same source of all that is." "How may we find wholeness in separateness?" In good and evil? As "catalysts moving us into new experiences of ourselves, without 'good' or 'bad'." Braden goes on to make the following points: "We are children of our Creator, however our Creator is perceived." "We have always been one, separated in our perceptions only." "Extremely evolved, skilled and masterful beings, any perceived limitations stem from our beliefs of what is possible, what exists, and what we are capable of." "We have never been separated from our source, our Creator, or one another . . . while experiencing as many possibilities of individualized expression."

"A life on Earth is an experience of choice." "Life is our path leading to our journey* home," Braden continues. "From this perspective, each expression of anger, pain and frustration betrays a place within seeking healing. We are expressions of the divine frequencies of love . . . perhaps our most valued tool will be the knowledge that we are not the pain, anger, or fear that we experience as we remember our truest nature."[17]

Similar to my own experience, Dr. Alberto Villoldo tell us, "All of us have internalized our culture's Judeo-Christian story of being cast out of paradise, when we were separated from our divine Creator." "When we let go of this story, we discover that we never left the Garden of Eden." "When we're in *ayni* (balance), paradise is our home; and physical, mental, and emotional health is our birthright. We discover that we never left Eden."[18]

"There are many books that tell us how to find God," writes Jason Shulman. "But the truth is that God is not lost or hiding. In fact, it is the actual, continuous, omnipresence of God that is so hard for the human mind to fathom." "God, Jesus, Amida, Buddha, Allah, enlightenment – whatever we call this Calling that calls to us from within our very body and from the world outside ourselves as well, always stands at the crossroads of every moment of life and death, offering us the answers to the great puzzle of being alive." "You have already knocked on the door of heaven many, many times." "*God, human, universe, home.*" "They are all One eventually."[19]

Jonathan Goldman, the expert in sound healing, writes, "Traditions (using the sacred vowel sounds) include the Enochi from Japan, Tibetan Buddhists, Islamic Sufis, Surat Shabd yoga practitioners, the Native American medicine people, and shamans of many indigenous cultures. Shamans, mystics, and sound healers throughout the world understand the power and importance of vowels." "According to Donald Beaman, the author of *The Tarot of Saqqara*, the Gnostics, a very early Christian sect, sounded the name of God as *IAO*."[20]

"Joscelyn Godwin, in *The Mystery of the Seven Vowels*, tells us that this understanding of the power and sacred energy of the vowels is said to have been known to the Egyptians, who then passed this knowledge on to the Greeks. He states that Nicomachus of Gerasa, 1st century AD, described a use of toning the seven sacred vowel sounds in order to act as 'the primary sounds emitted by the seven heavenly bodies.'" ". . . by repeating a mantra over and over, individuals are able to attune their own vibrational frequencies with the energy of this sound of power." Goldman writes he has "found that a mantra works almost as a kind of cosmic 'god' sound, where the reciter becomes like a celestial tuning fork that vibrates and attracts the energy of whatever or whomever one is sounding *to*." "Anytime a sound or word is repetitively used, it can be considered a mantra, whether it is an *om, shalom, Allah, Ave Maria, alleluia*, or *amen*." The vowel sounds of the Divine Name that Goldman tones on the CD that accompanies his book by that name can be considered a mantra.[21]

Alberto Villoldo was with his mentor/shaman friend, Antonio (don Jicaram), southeast of Cuzco, near the sacred Lake Titicaca and Puno and they had walked to the ancient city of Sillustani, home of the colossal round megalithic columns, monuments of perfectly cut and fitted granite blocks, and home of the ancient shamans of Peru. *Chullpas*, as these tombs of the caretakers of the Earth were called by the Ayamara, were towers made of blocks four feet by four feet, weighing at least ten tons. "This is *Umayo*" the shaman said. It was here in this sacred place that Antonio revealed another ceremony. ". . . for the Incas – the only great culture to develop south of the equator – the god-force is ascending. It rises from the Mother Earth. It rises from the Earth to the heavens like the golden corn. And those who are buried here at Sillustani are the men and women who spent their lives acquiring knowledge, germinating and cross-germinating their wisdom and their corn, discovering and understanding the forces of Nature and the relationship between the Sun and the Earth and the moon and the stars. The farmers perfected the alchemy with the soil and the shamans practiced the alchemy of the soul – not to produce *aurum vulgaris*, the common gold, but *aurum philosophus*."

"When the old ones who are buried here achieved mastery," Antonio continued, "when they knew what is was to be invisible, when they could influence the *past* as well as the future, and when they were able to keep a secret even from themselves, then the secret was revealed to them." "The ancient ones here knew that there was a relationship between time and light. That light has no time. Nothing can travel at the speed of light but light itself. If we approach the speed of light, we must become light. When we become light – an *Inca*, a child of the Sun – then time is dissolved." ". . . and who but those who understand that time turns like a wheel can manage to know the future and not let it upset their balance?" "Those who are buried here knew such things." "They understood the importance of life on Earth," he stated. "They knew that there would be no life on Earth if it were not for the Sun, that life is the direct result of their union." "The Sun is the father and the Earth is the mother and their parents were *One* – *Illa Tici Viracocha* – neither male nor female, energy in its purest form. They honored their mother and father by making *ayni* . . . a principle of reciprocity" (balance, mutual exchange, give and get in return). Antonio continued, "They say the shaman lives in perfect *ayni* – the universe reciprocates his every action, mirrors his intent back to him as he is a mirror to others. And that is why the shaman lives in synchronicity with Nature." ". . . I believe that *ayni* is always perfect, that our world is always a true reflection of our intent and our love and our actions."

It was here that Antonio taught Alberto the Prayer to the Four Winds – the South, *Amaru*, great serpent, ancient ones; the West, mother-sister jaguar, bridge between worlds, ancestral stewards of life; the North, the many-faced dragon, grandmothers and grandfathers; the East, *Aguilla Real*, great eagle, great seers and visionaries, mythmakers and storytellers; to Father Sun and Mother Earth; and to the Great Spirit, "know that all we do is in your name."[22] I use this basic form when I enter my little medicine wheel adding gratitude to my animal guide, the hummingbird. The Iroquois have a similar tradition adding *Up* to the Great Spirit/Father the Sky, *Down*, for Mother Earth, and *In*, with hand on heart.[23] Sometimes when we are trying to communicate with Spirit, God, Higher Self or guardian angels, thoughts of fear or chaos enter our frequency. Can you hear the angels saying, "Your signal is breaking up!" (like a cell phone)? "Please restore your connection!" "Come back to our wave-length."

In Section One, Part VII, of her book, *Letters from the Infinite*, Reverend Deborah Johnson shares what she received concerning the names of God: "I never said that I was the Father. That is your reference point for me. I said, 'I am the Living God,' 'I AM THAT I AM.' The names you give me reflect your societies' cultural values and norms. Some of your cultures describe me in terms of the elements, such as the wind, earth, or the sea. Some describe me as the energy force around them, such as life, evolution or transformation. Still others refer to me in terms of their own hierarchical structures, such as king, lord, master, or father. However, I am none of these things in the particular, yet I am all of them in the whole. The names you use to refer to me reflect what you hold in esteem and what you consider to be all-powerful."

"However, to see me in some things to the exclusion of others is to not see me at all. It is good that you see me in the opulent, but do you see me in the humble? It is fine that you see me in my majesty, but do you see me in my simplicity? I have no qualms about your referring to me as Lord, as long as it does not preclude you from seeing me in your mirror and in the face of your brother." "Do you see me in the gangs that you worry about, or even in the tax collectors? Not often enough."

"Remember that I cannot be captured in your words. I am the reference point for your words that transcends the words themselves." ". . . when you are stuck in seeing me in a particular form of expression, you fail to see me in all things. It is fine to see me as Order only if you can still see me in the midst of the chaos and claim my presence there. I neither come nor go. I am everywhere at all times. When you pray to me, I do not arrive. Your prayer opens your own eyes to receive my Good, which was merely awaiting your conscious recognition." "Since it says in Scripture that you were made in our image and likeness . . . a more linguistically correct translation of 'image' is 'nature.' You are made in a nature that is spiritual, and you have spiritual powers. When you know that you are One with me, you are able to direct these powers in a concentrated fashion for a Higher Purpose."

"The debate over whether I am to be represented in a male or female image is simply that – a figment of your imagination. I do not care what image you use, so long as it brings you closer in your understanding of your Oneness with me." "Yes, I am your father. I am also your mother, your sister, brother, friend, mate, partner, coach, confidante, shepherd, and the like. I am anything and everything that you need. It is important that you see me as all things so that you may see yourself that way." "I embrace all of you, individually and collectively. You are to do the same with and for each other."

"If you use your gender constructs to refer to me, the notion of being yin/yang would be more accurate. I am the fullness of both masculine and feminine energies; not merely one or the other. This is necessary, for I am Balance. It is necessary for you to recognize the same within yourself for you to be balanced also." "I am the roar of the ocean and the coolness of the breeze. I am spring in full blossom and winter in waiting." "In order to feel my love, you must see me as loving, forgiving, nurturing, embracing, and nonjudgmental." "I am Balance. I Am that I Am."[24]

So, you may ask, why are we talking about God, when this book is about going forward into the Fifth Sun of the Mayans or the Golden Age in a harmonious way? The Mayans definitely believed there was One Giver of Movement and Measure; they called this Creator – *Hunab K'u*. The present-day "Maya elders accord great reverence to the Heart of Heaven, which I believe is the . . . centre of our galaxy," says Patricia Mercier. She shares a Maya prayer:

> "May the Heart of the Universe be in my heart,
> May my heart be in the Heart of the Earth,
> May the Heart of the Earth be in my heart,
> May my heart be in the Heart of the Universe."[25]

She also quotes John Major Jenkins, "The solstice-galaxy alignment, (which) the Maya intended their 2012 date to target, is the end of humanity's descent into deepening illusion and confusion. We are about to turn the corner and begin an ascending 13,000-year cycle, toward a new golden age of light and truth revealed."[26]

Speusippus, one of Plato's 'old' Academics (the early school), defined God thus: "A being that lives immortality by means of Himself alone, sufficing for His own blessedness, the eternal Essence, cause of His own goodness." Rather male-sounding, but consider the times! "According to Plato, the *One* is the term most suitable for defining the Absolute, since the whole precedes the parts and diversity is dependent on unity, but unity not on diversity. The One, moreover, is before being, for *to be* is an attribute or condition of the One."[27] Other words used at that time were 'Divine Permanence' and 'unmoved Mover.' Huh? That Plato was really something, wasn't he?! As Herman Melville said, "Silence is the only Voice of God."

Neale Donald Walsch is one who has no qualms about the word 'God.' He wrote eight or nine books of dialogue he said he had with God, saying it was not writing so much as taking dictation; seven of his 27 books have made the *New York Times* bestsellers list. So there must be a lot of people who want to know about God. One of my favorites is *Happier Than God* because it has information that helps us better understand the Law of Attraction, which he calls the *Mechanism of Manifestation* or *The Process of Personal Creation*. Walsch refers to a bumper sticker from the sixties that asked, "IS GOD DEAD?" and writes, "I know, of course, that God is not dead. And most people agree with me. Surveys have shown that the largest percentage of people in every nation and in every culture believe in a power greater than themselves." He refers to the "Fast Track to Happiness" being taught today that indicates that we no longer need God; and, startlingly, he agrees! He does go on to explain, "How can we 'need' something that we always have, that we cannot *not* have under any circumstances, that we can always use, and that we cannot *not* use no matter how we might deny that we are?" "You cannot *not* have God in your life . . . this is something that many people cannot believe. They can't believe the highest promise of God: *I am always with you, even unto the end of time.*"

"They (also) can't accept the most wonderful truth taught by all religions, each in their own way: *Ask and you shall receive.*" Walsch continues, "All great truths begin as blasphemies. George Bernard Shaw famously said that, and he was right." "Take the idea that you can be happier than God." "Many people believe that life was meant to have a lot of pain in it." "It should be endured in silence. That earns you points in heaven." "When you talk about how you can be 'happier than God,' (some will tell you) you may be 'trafficking with the devil'."[28] Walsch asserts, "*Happiness is your natural state of being*, and you can occupy that space all of the time."

Then he asks, as many others have asked, "Where is God in this *Mechanism of Manifestation*? Is it all about cars, money and success?

Are those who are experiencing suffering to be blamed for 'incorrect thinking'?" (Sandra Anne Taylor speaks to these concerns as well in *Truth, Triumph, and Transformation*.) Walsch explains, "*Attraction* is a gift from a benevolent and compassionate deity. It is a tool with two handles – one in God's hands and one in ours." "Like all tools, it is most effective when it is used for the purpose for which it was intended." That purpose "is to create a happy, peaceful, joyous life for everyone whose life you touch, and for you, in that order." "This is the great gift of God: Continuous Power, Continuously On." ". . . Jesus said it directly and perfectly when he declared, 'As you believe, so will it be done unto you'." "The idea, as Eckhart Tolle made so brilliantly clear in *The Power of NOW*, is to *stay in the moment*. Don't 'futurize' and don't 'pasteurize.'" "Only when I get *out* of my past and stay *away* from my future can I genuinely experience what's happening right here, right now . . ."

Coming back to the names of God, Walsch explains, "If we use the word 'God' to indicate the Collective Divinity and the term 'you' to indicate the Individuation of Divinity, and if the day comes when you have raised your consciousness to a level where you are happier than many other Individuations . . . well, then, you will be *happier than 'God'*."[29]

Walsch defines two terms that I found very helpful. He speaks of the death experience or transition as *Continuation Day*. I like that. And *All That Is* means God *and* All Humans Everywhere (from his book, *Tomorrow's God*).[30] He also explains that nothing in life is predestined; we come into physical with something like a GPS device, "but you don't have to follow any particular route." "Whatever route you choose . . . the GPS simply makes an *instant adjustment*, creating a new way to get to where you (your soul) want to go *from where you are now*." ". . . there is no way you can get lost. And in that sense, *no way is the 'wrong way' to where you want to go*." ". . . we all wind up back Home with God in a Life That Never Ends . . ."[31]

"The *Upanishads*, the ancient scripture of India," writes Eckhart Tolle, "point to the . . . truth with these words: 'What cannot be seen with the eye, but that whereby the eye can see: know that alone to be Braham the Spirit and not what people here adore. What cannot be heard with the ear, but that whereby the can hear, know that alone to be Braham the Spirit . . . What cannot be thought with the mind, but that whereby the mind can think, know that alone to be Braham the Spirit . . .' God, the scripture is saying, is formless consciousness and the essence of who you are. Everything else is form, is 'what people here adore." Another word for Supreme Spirit used in Hinduism is *Atman* or the Indwelling God.

Tolle tells us, "The inspiration for the title of this book (*A New Earth*) came from a Bible prophecy that seems more applicable now than at any other time in history." ". . . both the Old and the New Testament speak of the collapse of the existing world order and the arising of 'a new heaven and a new earth.' (Matthew 5:3)" "We need to understand that heaven is not a location but refers to the inner realm of consciousness." "Earth . . . is the outer manifestation in form, which is always a reflection of the inner. Collective human consciousness and life on our planet are intrinsically connected. *A new heaven' is the emergence of a transformed state of human consciousness, and 'a new earth' is its reflection in the physical realm.* Since human life and human consciousness are intrinsically one with the life of the planet, as the old consciousness dissolves, there are bound to be synchronistic geographic and climatic natural upheavals in many parts of the planet, some of which we are already witnessing now."[32] When we think of the implied unity in duality of 'As above, so below,' we want to bring them together as in bringing the attributes of heaven (the spirit) and combining them with the wonderful characteristics of earth (the physical life) – oneness or wholeness. Can we bring our ego mind and our Higher Self in closer communion?

Of course, bringing heaven to Earth was the primary vision of Maharishi Mahesh Yogi, believing that contacting our essential Self in the pure awareness of transcendental meditation would be instrumental in doing this. If you lean more toward scientific proof, Amit Goswami, Ph.D. wrote a very thorough explanation in *God is Not Dead: What Quantum Physics Tells Us about Our Origins and How We Should Live*. Some call God the *Origin*, or *The Comforter, The Great I Am, The Great Spirit, The Unthinkable*; others say God is Love. Meister Eckhart said, "God is novissimus" – the newest thing in the universe. That one throws our minds into a spin until you think of the Bible quote, "I am the Alpha and the Omega" – the beginning and the end.

The *Wiccan* tradition has its roots in the North European Druids, Celtic beliefs, and the Saxons and like the Mayans and other indigenous peoples worshipped Nature, Mother Earth and Father Sun, and the four elements of Earth, Wind, Fire and Water. The equality of God and Goddess sets Wicca apart from the Judeo-Christian and Islamic beliefs. Wiccacraft means the craft of the wise, much like the shamans we have talked about earlier who were also adept at tuning in to the rhythms of nature. A peripheral part of Wicca, Celtic and Druidism is the belief that certain trees have healing powers, as elaborated in *The Healing Power of Trees* (Llewellyn Publications) by Sharlyn Hidalgo reviewed by One Spirit Book Club (www.onespirit.com). She writes, "Trees . . . stand between Mother Earth and Father Sky as we do. We are both grand bridges between these two spheres of consciousness."

In *Natural Grace* we find the four dimensions of prayer, as described by Matthew Fox and Rupert Sheldrake: "#1. Awe and praise, #2. Silence, letting go and emptying, #3. Entering the creative process by honoring our desires, joy and sorrows, and #4. Transformation by compassion, realization of our interdependence." They quote Meister Eckhart, "Be still, be empty, be with being, be with non-being, be with nothingness. . ." "If the only prayer you say in your whole life is 'thank you,' that would suffice."[33]

The Lakota pray, *"Aho Mitakuye Oyasin"* – 'all our relations, we are all connected' at the end of each morning prayer.

A prayer from Ute (Native American tribe) Wisdom:

> Earth teach me humility
> As blossoms are humble with beginning
> Earth teach me courage
> As the tree which stands alone
> Earth teach me regeneration
> As the seed which rises in the spring
> Earth teach me to forget myself
> As the melted snow forgets its life
> Earth teach me to remember kindness
> As dry fields weep in the rain

Mike Dooley speaks of the All-That-Is as the Universe or Divine Intelligence and that seems to include much more than the physical universe. He writes, "It comes from a 'place' or dimension that 'precedes' both time and space, and from 'there' it has created their intersection – this platform from which material manifestations exist." "For our . . . Earth to have such dazzling colors, sounds, and exquisite beauty, it shouldn't take much of a leap to consider that the wellspring* of such a paradise would be equally – if not far more – radiant and spectacular." "It's endowed with intelligence, and it's what you've been taught is 'God,' yet it exists without any religious trappings, judgments, and rules. Every grain of sand, the trees, the air, the water – everything, including all that is unseen – possesses this intelligence, this awareness . . . an awareness that includes you and your thoughts." He further relates, "I'm not a big proponent for using the word 'prayer.' It's like using the word, 'God' in that it has so many connotations, meaning vastly different things to different people. ". . . it usually implies that the one doing the praying is powerless and beholden to powers, or to a god – outside themselves – who picks and chooses whom he will help and when. But I love prayer's implication that we are not alone and that we are being heard." "*You are never alone.*"[34]

Again, from *Woman Prayers*:

> Earth is crammed with heaven,
> And every common bush afire with God,
> but only he who sees
> takes off his shoes.
> Elizabeth Barrett Browning
> England, 1806-1861

also,

> When we do the best we can,
> We never know what miracles await.
> Helen Keller, 1880-1968[35]

Caroline Myss has been in the field of energy medicine and human consciousness for over 20 years (from her book jacket), but as she writes in *Defy Gravity*, "In the early years of my work as a medical intuitive, I . . . clung to the professions of 'writer' and 'teacher' as if they were designer labels." ". . . I never had to explain (those occupations), whereas describing myself as a medical intuitive always required a lengthy and exhausting description." "But this rice-paper-thin wall between health and illness, life and death . . . is exactly the wall I had successfully avoided until people began to experience healings during workshops . . . based on (my book) *Entering the Castle* . . . a contemporary view of the classical mystical experience (based on) . . . the magnificent teaching of the 16th century mystic. . . Saint Teresa of Ávila." "As I studied Teresa's writings, I realized the emptiness that people continually express today. . ." "People are missing a sense of awe in their lives, a connection to the sacred. . ." "They don't want to talk about God; they want to feel the power of God."[36]

Myss continues, ". . .I direct readers to leave their reason at the door, so to speak, and enter the realm of mystical consciousness – not just for healing but as a way of life. . ." So this book is based on leaving 'serious' and 'weighty' matters, and as a mystic, "defy gravity." "A mystic 'perceives' life through the eyes of the soul, (one) who experiences the power of God rather than speaks or debates the politics of God." Myss asks us to "Consider that we are living at a major turning point in the history of humanity, a time of great crisis and great opportunity. It may be unreasonable to imagine that you can make a difference to a world in crisis based on how you undertake your own healing, yet I believe this to be true. It makes no sense to our logical minds that as we heal, the whole of life heals. Yet the power of one mustard seed can move a mountain; the power of one clear light does illuminate the darkness; the power of a person devoted to truth becomes a channel for healing grace that benefits all humanity. No matter what you heal within yourself, from a negative thought to a progressive cancer, the very act of healing has in some way made a difference to everyone on the planet. It is a truth that goes beyond the bounds of ordinary reason, but so does all mystical truth. . ." ". . . Jesus' message to us (was) to embody a level of consciousness that transcended human reason: . . . a higher law that rules the spirit, a mystical law that holds no allegiance to the laws of religion." ". . . Jesus insisted that the presence of *Abba*, as he called the father-God in his native Aramaic, is a force of love so powerful it allows you to trust that beyond whatever you may endure lie greater cosmic reasons for your experiences." Myss concludes, "Live as though you have the power to change the world – because you do."[37]

Echo Bodine, gifted psychic, writer and teacher, writes about her life: "As time passed my concept of God completely changed. I saw in meditation that the light . . . at the core of my being was God, and I realized what the expression, 'I am God' meant . . . I am simply affirming that God is the core of me. God expresses through me. I *am* one with God." "This is my foundation, my inner strength and knowingness." "Over the course of many years deeper understanding gradually came." "When I really understood that it wasn't God creating the pain in my life, but instead those experiences were life lessons my soul chose, my prayers changed and my relationship to God continued to change and heal."[38]

Rev. Michael Bernard Beckwith writes, "As we mature around our concept of the Godhead, we take responsibility for our lives. *We grow out of our childhood fantasies that there is a Great Something outside of us manipulating the environment, running the affairs of the universe by a reward-punishment system. We give up our status as a beggar and let God off the hook of being Santa Claus.*"[39]

You are not 'taking away God's power' (which is impossible anyway!) when you become a conscious co-creator. The Creator made you in that image – a creator. You are creating your life with every thought and belief; now that we understand the process, we can create more to our liking. Let's look at the way Love can help us with that.

Love and Heaven on Earth

In *Infinite Possibilities* Mike Dooley writes, "I believe there are two kinds of love: the one that people talk about when they describe falling in love and the kind that God or the Universe, possesses for us. These two kinds of love are actually so different that there should really be a whole new word to describe the kind that the Universe exudes. But since there's no such word, I'll call it 'Infinite Love'." "When considering the magnitude of the Universe and the perfection, balance, and harmony that exists in nature, we're obviously dealing with a creator made wholly of Infinite Love." "Thus, its creations and all that comes from it could only ever be of Infinite Love."

He continues, "*Infinite Love is everywhere, always.* It *is* time, space, and matter. It *is* thought, awareness, and energy. It's the present moment and every form of consciousness it contains. It's the desire and fulfillment of every dream ever dreamt, the spirit of life in an endless dance of joyful becoming. This love . . . can't be escaped . . . ; it's absolute. Right now, no matter who you are, it holds you in the palm of its hand – a hand big enough to hold, keep, and *understand* the most misguided and vile of all people. This kind of Infinite Love is not an emotional love; it's not contingent on time, space, or matter; thoughts, beliefs, or perceptions." ". . . for the pockets in the world where you don't see evidence of this Infinite Love, perhaps it's because we can't see all things from the extremely limited vantage point offered by our physical senses alone."[40]

Gregg Braden dedicates *The Isaiah Effect* in this way: "This book is dedicated to our search for love and the memory of our power to bring heaven to earth." He writes about love from the view of the Essenes, the scholars of the Qumran communities, whose writings predate the Christian

Bible by several hundred years, "From their (the Essenes) perspective, we are one with the Father in heaven." "To focus our prayer, we must love the creative principle of life itself, our Creator, with all of our heart, soul, mind and strength. *Because we are one with our Father in heaven, in doing so, we have just loved ourselves.*"[41]

Gill Edwards, psychologist and author from the UK, says it very clearly, "We do not have to earn or deserve love. We do not have to 'behave well' or conform to external rules and expectations. In a loving universe, we can relax. We are safe. We are worthy. We are loved without condition. We are cosmic voyagers on a magnificent adventure in physical reality, and – as creative sparks of the divine – we can have, do, or be anything we wish. No limits. No strings attached. We can create our own heaven on Earth. And the key to doing so is (having) unconditional love – for self, others, and the world."[42]

If you have not read Dr. Masaru Emoto's first book, *The Hidden Messages in Water*, it explains his research with ice crystals formed from water exposed to various thoughts, words (spoken and written) and pollutants. He then photographed the resulting ice crystals and found something amazing: those exposed to positive, caring vibrations made beautiful, symmetrical crystals, while those exposed to negativity and pollution made lop-sided, irregular incomplete shapes. "Water exposed to love and thanks sparked the most beautiful response." His research was upon the vibratory or magnetic resonance of water, called *hado* in his language. (Japanese)

Lynn McTaggart (whom we heard from earlier in *The Intention Experiment*) also reported on similar tests carried out by Dean Radin, who confirmed, "Emoto claims to have carried out hundreds of tests showing that even a single word of positive intent or negative intent profoundly changes the water's internal organization. The water subjected to the positive intent supposedly develops a beautiful, complex crystalline structure when frozen, whereas the structure of water exposed to negative emotions became random, disordered, even grotesque. The most positive results supposedly occur with feelings of love or gratitude." "A statistically significant number of the . . . judges concluded that water sent the positive intentions had formed the more aesthetically pleasing crystalline structure."[43] ". . .Japanese for 'I love you' (is) "I *am loving* you." Emoto writes, "In love, agape (unconditional, self-sacrificing love) is often compared with eros (passionate love with sensual desire). I believe that eros is changeable while agape is not. Thus eros can generate more energy. When eros changes into agape, vast love will wrap you in gentle warmth." Emoto, who lives in Tokyo, reasoned that since the human body is 70% water, the vibration of thought, the vibration of love has a strong effect. I especially liked his experiment of exposing water to the song "Imagine" (one of my favorites) composed by John Lennon, and found that the resulting crystal was "beautiful and dreamy-looking."[44] (See also chapter on 'Global Issues")

Marc Ian Barasch writes, "In Judaism, the *hesed* ('steadfast love') refers to God's unbounded caring that never wavers – as well as what those who are made in His image are capable of reflecting. *Agape*, an ancient Greek word for the most unlimited, unselfish, accepting form of love, was adopted by early Christians to describe humanity's (nearest) attainment of God's love for His creation. . . 'a feel for the preciousness of all human beings.'" "Here compassion is synonymous with a love not meted out according to merit, but falling like spring rain on the virtuous and the sinner alike." "Hindu spiritual teacher Neem Karoli Baba said, "What astounded me when I was around Maharaji wasn't* that he loved everybody . . . (but that) I loved everybody. It's a kind of love that's contagious." He further illuminates in *Field Notes on the Compassionate Life*, "Charity extends the privileges of insiders to outsiders; while *agape* erases the line between insiders and outsiders entirely."[45] Love erases the duality.

While on the subject of agape, let's see what Michael Bernard Beckwith, founder of the Agape International Spiritual Center and featured teacher in *The Secret* has to say: "Today's consciousness research and quantum physics reveal that consciousness interpenetrates all that is, affirming that existence is not random, that it is intentional and governed by universal law. To that I add love. We are governed by law and love. There is no absolute definition of God or Love. They are synonymously indefinable. However, we intrinsically know when we experience a burst of realization or sense a presence that can only be called God or Love. Neither is visible, nor can they be analyzed by the thinking mind or intellect. They can only be realized. "Although I use such words as Source, Love-Beauty, Presence, Ineffable, and Spirit to represent the Godhead, none of them is adequate to describe the Great Something we call God that is everywhere in its fullness."

I love the new word Beckwith created: 'Blissipline' which emphasizes that *"Discipline is a practice of self-love, self-respect, and surrender that results in freedom."* Or from one of his and his wife, Rickie's songs, "Trust love. Trust God that love is everywhere, that we are here to be perfect givers of love and receivers of love. Love will have the final word."[46] Thomas Merton (1915-1968), the Trappist monk of the Abbey of Gethsemani in Kentucky, wrote *Choosing to Love the World* and the title gives us a message: Love is a Choice. Or as I like to say, "Love it all. Love the world into change. The 'bottom line' is always love!"

Freddy Silva says he finds Love in crop circles. Pulitzer Prize winning author, Willa Cather (1873-1947) said, "Where there is great love there are always miracles." James Redfield wrote, "Love, of course, is the best-known measure of inner transcendence." "It is a love that exists without an intended focus, and it becomes a pervasive constant that keeps our other emotions in perspective." (From *The Celestine Vision*)[47] In *God and the Evolving Universe,* he and co-authors state, "*Agape* can transform all relationships." "How deep are our spiritual experiences? The level of our love is our best gauge of this." "Love grows through intention and practice."[48]

". . . there is only one thing I ever work on with anyone," writes Louise L. Hay, "and this is *Loving the Self.* Love is the miracle cure. Loving ourselves works miracles in our lives." On the subject of overweight, she advises, "When we begin to love and approve of ourselves, it's amazing how weight just disappears from our bodies." "*In the Spiritual Realm* there is prayer, there is meditation, and becoming connected to your Higher Source. Practicing forgiveness and unconditional love to me are spiritual practices." On the subject of prosperity Hay says, "Love Your Bills!" "The creditor assumes you are affluent enough to and gives you the service or product first." "If you pay with love and joy, you open the freeflowing channel of abundance."[49] Great advice.

Of course, the greatest 'treatise' on love comes from the Apostle Paul, I Corinthians 13:4-6 "Love is patient; love is kind and envies no one. Love is never boastful, nor conceited, nor rude; never selfish, not quick to take offence. Love keeps no score of wrongs; does not gloat over other (person's) sins, but delights in the truth." And verse 13: "In a word, there are three things that last forever: faith, hope and love; but the greatest of them all is love."[50]

Gregg Braden shares a quote adapted from *Meditations With the Hopi*, by Robert Boissiere (Bear & Company): "Give love to all things, mountains, trees and rocks for the spirit is one though the kachinas are many." He also writes, "Compassion is the kernel of your very nature." And he gives us the formula:

"AS
we allow life to show us ourselves in new ways
so that we may know ourselves in those ways
AND
we reconcile within ourselves that which life has shown us
THEN
we become Compassion." [51]

"We can see in our daily experience that there are, in fact, many people who carry in their presence a magnanimity and life-enhancing energetic that invites tolerance, spaciousness for difference, and a capacity to be comfortable with ambiguity. When we look closely, we find a story unfolding within consciousness itself that is subtly taking shape and transforming our world from the inside out," writes James O'Dea. There is ". . . much more love present in the world than is reflected in the dominant institutions or media." "You are coming to know, in a primary, experiential way, that love is at the heart of evolution and the source of healing. Love is an evolutionary force because it is the primogenitor of nurturance." ". . . it is to be understood . . . as a quickening in the center of our being – a quickening less susceptible to fear arousal . . . but stimulated by experiences of wholeness, unity, and interconnection. It is what philosophers and theologians call the emergence of nondual consciousness." O'Dea continues, "Our collective conscious has been evolving." "Many indigenous people think of this juncture in history – and for some, 2012 in particular – as a time of cleansing." "But for the indigenous, Nature is not something other than human life; it is not to be conceived as a separate force coming at the end of time to judge and punish us for our sinful addictions to unsustainable progress. We are a part of Nature. . ." [52]

How Do We Get There From Here?

We have talked about some pretty 'heavy stuff' in these pages. If, as we have seen, each of us can help raise the collective consciousness that will bring in a new age of hope and harmony, how do we do that? How can we each raise our level of consciousness? First, I must ask you what you love. What makes you happy? Do you remember when you were a kid and someone asked what your favorite color or favorite animal was? Have you made one of those lists lately? I have found myself at several points in my life in one stage of depression or another. I have had psychological help; I have taken medications, but the day to day work on changing my thought patterns has turned out to be the most effective way out – when I remember to use the tools I have! My list of *'Loves'* is one of my tools. My husband, daughter, granddaughter, The All-That-Is/Creator, popcorn and crunchy foods (!), sunsets, hummingbirds, thunder, and waterfalls are just a few of the items on my list of forty or more. There are activities I like such as sitting by the Pacific, reading, gardening, photography and making bouquets. One of my favorite places, Mt. Wilson and the Meadows that lie below it, are on that list. I have had a book in process for some time called *My Love Affair with a Mountain*, and just thinking about that mountain makes me happy. I have a need to express the love I feel for Nature:

> The mountains lift me up
> The ocean cleanses my soul
> Cliffs remind me beauty is everywhere
> Valleys show me how to flow in peace

2013

> The sky opens my heart
> The ancient trees teach me patience
> And the clouds draw me upward
> with their beauty

I have always said, "There are as many paths to God or Higher Consciousness as there are people on this Earth," and that amounts to about seven billion now. If you love charity and volunteer work, there are so many opportunities, from helping in a Soup Kitchen to running to raise money and awareness for cancer research. If you love Nature and being outdoors enjoying that, some of those activities may be your route to happiness and joy, leading you to an awareness of your inner self. Reading has been one of my paths; I have felt for over 30 years that I have been led to the next book or subject that I needed in order to grow. I'm sure I must have asked on some level for that help from the Universe. Creating your own list of 'Loves' can lift you up.

If you lean toward improving the physical body, getting in the 'Zone' may be your course. Many people find hiking, bicycling, or running their preferred method of 'being with' their inner Self. Author, Bill Plotkin, a university-based psychologist, professor and research scientist, left it all to wander on foot in the Pacific Northwest for a year and a half. After a brief interim he says, "I heard the soul's call again, this time as an urgency for wilderness solitude, to look inwardly as far and as innocently as possible, and to wait until some truth rang out," he writes. "I would go out on the land, alone, to a wild place, dwell there for several days without food, look into the mirror of nature, and cry for a vision." I'm sure many of you have heard of the 'vision quest' that has been copied from the Native American traditions; it is offered by many teachers. Plotkin continues, "Each of us is born with a treasure, an essence, a seed of quiescent potential, secreted for safekeeping in the center of our being. This treasure, this personal quality, power, talent, or gift (or set of such qualities), is ours to develop, embody, and offer to our communities. . ." "Our personal destiny is to *become* that treasure through our actions." He explains that, ". . . these three stages of the journey* - severance (temporary or permanent, especially from your way of understanding yourself), initiation (into the life of your soul), and incorporation (into a new role, a new place in your community) –

correspond to the three stages of any rite of passage." "You will have to move beyond any requirement to be 'good,' any obligation to be held back by shame about who you are." He quotes Mary Oliver from her poem "Wild Geese":

> You do not have to be good.
> You do not have to walk on your knees
> for a hundred miles through the desert, repenting.
> You only have to let the soft animal of your body
> love what it loves.[53]

Plotkin advises, ". . . like most encounters with the Great Mystery, a call to adventure is typically experienced as uncanny or numinous, suffused with the sacred or holy. More often than not it is accompanied by enormously powerful emotion. . ." "Nonordinary states of consciousness are also common, states in which you may apprehend something astonishing about the world for the first time." He concludes, "When a sufficient number of contemporary people have reentered nature's soulstream and become conscious contributors to the unfolding story of the world, industrialized nations might mature into sustainable, ecocentric, and soulcentric communities, inhabited by people who are wildly creative, imaginative, adventurous, tolerant, generous, joyous, and cooperative members of the more-than-human world. This is my prayer."[54] I would say his decision to 'follow his bliss' must have been guided by his Inner Self; a great book and a new career came out of it.

In the Foreword to Plotkin's book, *Soulcraft*, Thomas Berry writes, "Further, in our association with indigenous peoples, we began to appreciate the profound sense of realism they manifested in the ritual communion of the human soul with the deeper powers of the universe. In these early cultures, the universe was experienced primarily as a presence *to be communed with and instructed by*, not a collection of natural resources *to be used* for utilitarian purposes. The winds, the mountains, the soaring birds, the wildlife roaming the forests, the stars splashed across the heavens in the dark of night; these were all communicating the deepest experiences that humans would ever know."[55] Reiki and yoga are two other methods of finding soul through body work.

This brings up another *Path*: the study of ancient and modern indigenous peoples and/or the various tribes of Native Americans here in the U.S. I felt drawn to the Mayans; you might find resonance with the Maoris. The Incan tradition, especially its shamanism, came to be the focus of Alberto Villoldo's life. His sharing about his friendship with shaman don Jicaram and the four levels of consciousness enriched my own journey. Because of his books I was aware of the symbols associated with these levels before I sought help from Dr. Stephen Banko, shaman of a similar tradition in North Carolina. Knowing that serpent, jaguar, hummingbird and eagle were levels of understanding, my being gifted during this healing by the 'ancestors' of the personal symbol of the *hummingbird* was very humbling and gratifying. An exercise that Villoldo describes in *Courageous Dreaming* of 'Burning Our Roles' is a powerful ritual and potent psychological application. "On each strip of paper, write a role, label or self-definition that you identify with. Be sure to include *husband, wife, father, mother, doctor, breadwinner, nurse, recovering alcoholic, student, or lover*" (whatever applies to YOU) and so on. "Wrap each strip around a twig, and thank each role for the lessons it has taught you and the powers it has bestowed upon you. Bless each role, and then place the twig in the fire and watch it burn . . . know that you're creating a sacred ritual for yourself without engaging the mind." "Know that you cannot be defined by your roles, but you can perform them all with beauty and grace."[56] So many paths.

You may want to start with clearing some 'stuff' with traditional psychotherapy, spiritual therapy, past life regression, or even help from a shaman. Perhaps you are on the other side of this coin and helping others find their authentic self is your greatest fulfillment. Early in my search I was advised by a psychotherapist to begin writing *affirmations*. Are you familiar with affirmations? The dictionary defines 'affirmation' as confirmation, ratification, assertion; the thesaurus adds allegation, acknowledgment, declaration, and oath. I might add: A statement of my truth about myself, my life and my world. Affirmations are a powerful path on the journey. Writing these was therapeutic and I wrote a book (unpublished, but helpful to me) of those that I needed to more fully establish in my mind, and illustrated it with my most beautiful photographs. I included "*The Universe loves me,*" (which I did not really believe at the time, but wanted to) and "*I see beauty in everything,*" thoughts that I needed to bring into my consciousness and belief system.

Louise Hay writes, "An affirmation is a beginning point. It opens the way. You are saying to your subconscious mind: "I am taking responsibility." "I am aware there is something I can do to change." If you continue to say the affirmation, either you will be ready to let whatever it is go, and the affirmation will become true; (for you)* or it will open a new avenue to you." "You will be led to the next step that will help you with your healing." "The moment you say affirmations, you are stepping out of the victim role. You are no longer helpless. You are acknowledging your own power." Her book, *Heart Thoughts*, takes you from A to W, from Abundance to World Community with affirmations, meditations and excerpts on many, many subjects. She guides us toward connecting to our Inner Self with unconditional love.[57]

I am not an avid journal keeper, which is another helpful tool, but I do enjoy writing. If you are reading this you will know that I have gone from 'enjoying' to 'sharing.' When I was going through a divorce, I began writing *Sunset – Sunrise* as a way to work through my feelings around that part of my life, and continued the book after my second husband died. If we listen to our inner self, we will make a path, even if it takes a machete! The work of Abraham-Hicks is very much about helping us stay in touch with our Inner Being and has helped me come to a better 'place.'

Along this same track, I write poetry and love to read poetry. Seeing in *black-and-white* the thoughts and feelings of others can help us get in touch with our own. Roger Housden's compilations of poetry, especially *ten poems to change your life, ten poems to last a lifetime, ten poems to open your heart,* and *ten poems to set you free,* are wonderful. From the Introduction to that last book he quotes the beginning line from a poem by Miguel de Unamuno:

"*Shake off this sadness, and recover your spirit*

Unamuno's message is a call to action, and when it comes to kindling the fire in your life, action is often, though not always, what is needed. Remember those lines by Mary Oliver that begins her poem, "The Journey":

> *One day you finally knew*
> *what you had to do, and began,"*

And as Rumi says in Coleman Bark's version called "Unfold Your Own Myth,":

> *But don't be satisfied with stories, how things*
> *have gone with others. Unfold*
> *your own myth, without complicated explanations, . . .*[58]

Housden writes in *ten poems to open your heart*, ". . . the poems are a tribute to the resonance that lives in you, the reader, who has picked up this book. You, who know love when you feel it – who know, along with Rumi, that

> *All the particles in the world*
> *Are in love and looking for lovers.*
> *Pieces of straw tremble*
> *In the presence of amber.*[59]

Saved by a Poem by Kim Rosen is another book you might enjoy. She quotes one by Derek Walcott:

> *Give back your heart*
> *to itself, to the stranger who has loved you*
>
> *all your life, whom you ignored*
> *for another, who knows you by heart.*[60]

Daniel Ladinsky has translated and written three books of the poems of the Great Sufi Master, Hafiz, plus *Love Poems from God: Twelve Sacred Voices from the East and West*, one voice of which is, of course, Hafiz. He writes, "Millions of people in this world believe that Rumi (1207-1273), St. Francis (1182-1226), Hafiz (1320-1389), Kabir (1440-1518), and Tukaram (1608-1649, born in Maharashta, India) became divinely wed to the Infinite, achieved union with God, and thus became aspects of

His voice." Rabia (717-801AD) of Basra, Mesopotamia was an influential female Islamic saint who wrote with sensuousness and, at times, "graphic eroticism," writes Ladinsky. He quotes from her collection:

The Perfect Stillness

Love is
the perfect stillness
and the greatest excitement, the most profound act,
and the word almost as complete
as His name.

He also quotes Rumi:

He Asked for Charity

God came to my house and asked for charity
And I fell on my knees and
cried, "Beloved,

what may I
give?

"Just love," He said.
"Just love."

I could go on and on – but, perhaps, you would like to read the book. Just one last quote from Tukaram:

I Might Act Serious

If God would stop telling jokes,
I might act
serious.[61]

Some people choose the path of devotion, as did these poets. Prayer was their life, it seems. Religion is not my 'strong suit' and that subject has been 'thoroughly hashed out' in other places. However, I *did* find in a wonderful book, *Opening to You: Zen-Inspired translations of the PSALMS*, by Zen priest and poet, Norman Fischer. He spent a week at the Gethsemani Abbey, Thomas Merton's monastery (mentioned above) and found the chanting of the Psalms by the monks to be very different from that which he grew up with in Hebrew. He says, "I saw that these good brothers of Gethsemani were true treaders of the path, sincere practitioners, possessed of wisdom and knowledge. If the Psalms had meaning for them, clearly I was missing something. I felt I had to investigate for myself." "My effort has been to bring out their latent meanings, rather than create new meanings." "I call them 'Zen-inspired' because I approach them the only way I can: as a Zen practitioner and teacher, with a Zen eye." "Buddhism begins with suffering and the end of suffering and the path toward the end of suffering." "The Psalms make it clear that suffering is not to be escaped or bypassed. Much to the contrary, suffering returns again and again, a path in itself, and through the very suffering and the admission of suffering, the letting go of suffering and the calling out from it, mercy and peace can come. . ." (*Abraham* in the Abraham-Hicks materials® uses the word 'contrast' for suffering.) "The Psalms are poems." "Making language is making prayer. Our utterances, whether silent or voiced, written or thought, distinct or vague, repeated or fleeting, are always essentially prayer, even though we seldom realize it," relates Fischer. "In the end prayer is not some specialized religious exercise: it is just what comes out of our mouths if we truly pay attention. Debased as it so often is, language at its core always springs forth from what is fundamental in the human heart."[62]

"Though I realize that the idea of seeing God as *you* isn't unique (it is a common trope in medieval Sufi poetry), it had a very personal, almost private dimension for me," writes Fischer. "Poetry evokes the unknowable." "Since a cardinal principle of Buddhist thought is precisely that it be nontheistic,* there has been continuous criticism of such doctrines, contending that they are . . . attempts to introduce the concept of God into the Buddhist system of thought." "The Psalms are historical documents of a particular people. . ." Fischer concludes this explanation by saying, "I have come to think, after working for many years intimately with many people along the course of their heartfelt spiritual journeys,* that traditions now need to listen to the human heart before them . . . more than they listen to their various doctrines and beliefs." With this in mind I share his version of

Psalms 23:

"You are my shepherd, I am content
You lead me to rest in the sweet grasses
To lie down by the quiet waters
And I am refreshed

You lead me down the right path
The path that unwinds in the patter of your name

And even if I walk through the valley of the shadow of death
I will not fear
For you are with me
Comforting me with your rod and staff
Showing me each step

You prepare a table for me
In the midst of my adversity
And moisten my head with oil

Surely my cup is overflowing
And goodness and kindness will follow me
All the days of my life
And in the long days beyond
I will always live within your house"[63]

If ever you feel sad or 'down,' remember there is a lot out there to lift us up. *Woman Prayers* is a collection of prayers and poetry by women 'throughout history, around the world,' as the title says. One traditional Navajo chant there:

"The mountain,
I am part of it. . .
The herbs, the fir tree,
I am part of it.
The morning mists, the clouds, the gathering waters,
I am part of it.
The wilderness, the dew drops, the pollen. . .
I am part of it."

And a Traditional Buddhist Prayer is in this book:

"May I be free from danger,
May I be free from fear,
May I be healthy
May I dwell in peace.

May you be free from danger,
May you be free from fear,
May you be healthy,
May you dwell in peace.

May all beings be free from danger,
May all beings be free from fear,
May all beings be healthy,
May all beings dwell in peace."

Another Native American prayer, very simple, is called "Unexpected Blessings"

Give thanks
For unknown blessings
Already on their way."[64]

Over the years I have written enough poetry to fill a small book and have had some published in the literary publication of Fort Lewis College. I find I have to be in one of three emotional 'spaces' to write – sad, distraught or ecstatic. I will share here a poem called "Morning Prayer" that I wrote several years ago.

> Thank you, Universal One,
> The All, Mother/Father, Creator
> For this another day of life
> In this Incarnation.
> Help me to see it as a gift,
> A miraculous present,
> A package I may unwrap and enjoy.
>
> Help me live in this moment,
> tossing away all the garbage from the past
> and knowing the future is only
> an illusion.
> Teach me moment to moment awareness,
> Minute by minute renewal,
> Gratitude in this moment.
>
> Teach me to live with joy,
> even in the midst of pain.
> Help me to be grateful
> even in sorrow.
> Help me to blossom into wholeness,
> into the fullness of my potential.
>
> May I always remember that
> the greatest law is the Law of Love.
> When in doubt – love.
> When angry – love.
> When scared – love.
> Love it all.
> What did you think needed love?
>
> Help me to have faith
> in the perfection of all

when all is dark.
And remember Grace is there
if I but call out for it.

Teach me to see the perfection
of every moment
wherever I am,
and the miracle in everything.
Knowing that
Expecting a miracle
creates a miracle.
And bliss is attainable.

Thank you God for this day.

Another great book on this path of devotion is *Women Pray: Voices through the Ages, from Many Faiths, Cultures, and Traditions* edited and with an introduction by Monica Furlong.[65] The path of prayer and devotion brings to mind meditation. We have talked about Transcendental Meditation in the chapter on Consciousness and for more years than I care to disclose have found it to be my anchor. Though over time, the practice has evolved as I have learned and tried other types of mantras, etc. Most recently the Abraham-Hicks *Getting Into the Vortex Guided Meditation CD and User Guide*[66] has been a great addition to my 'tools for the journey.' Because your 'frequency' is unique, your meditation will evolve and become 'yours,' your style, your 'zone.'

A meditation technique I use frequently comes from Dharma Singh Khalsa, M.D. in his book (with Cameron Stauth) *Meditation as Medicine*. He was a licensed anesthesiologist before being accepted by Yogi Bhajan into the 'Brotherhood of the Khalsa' meaning Pure One, which is the Sikhism form of the yoga traditions. In this practice one repeats, silently, the mantra "*Sa ta na ma*" while touching in turn, with each syllable, the forefinger, middle finger, ring finger and little finger to your thumb, on both hands. I breathe in with one 'round' and out on the next. There is a full explanation and instructions in the book. Dr. Khalsa gives examples of how patients have been helped and healed by his methods as well as a chart of the chakras and what parts and functions they each have in the body.[67]

Along a more Christian outlook is Hugh Prather's *Morning Notes: 365 Meditations to Wake You Up* (Conari Press). For those who can't seem to find time there is a book titled *20-Minute Retreats: Revive Your Spirits in Just Minutes a Day with Simple Self-Led Exercises* by Rachel Harris, Ph.D. (Henry Holt and Company, LLC) I recently heard about The Universal Life Church, which is non-denominational. There is a path for each person. There are many seekers who are closely following the prophecies that came from the appearance of The Blessed Virgin Mary and the Archangel Michael to four schoolgirls in Garabandal in northern Spain from 1961 to 1965. It has been reported that there will be an 8-day warning and a time of purification; everyone on earth will know that the time has come to choose a spiritual life or worldly life. Each must see him/herself for what they truly are; this is the admonition given.

The path of Nature appreciation is another of my favorites. I love the mountains and scenic photography. Alberto Villoldo's teacher/shaman told him, "The *apus*, the great mountain peaks, give you strength to endure your work; the heavens give you harmony."[68] I can understand this completely. I mentioned Mt. Wilson in my list of 'Loves,' and I try to spend several days alone with 'my mountain,' each summer. It's wonderful when you can find a 'spot' that nourishes your soul and spend some quiet time there. Nancy Wood brings my love of poetry and mountains together in her poem "Of Mountains and Women."

> The hearts of mountains
> and the hearts of women
> Are both the same. They beat
> an old rhythm, an old song.
>
> Mountains and women
> are made from the sinew of the rock.
> Mountains and women
> are home to the spirits of the earth.
> Mountains and women
> are created with beauty all around

> Mountains and women
> embrace the mystery of life.
>
> Mountains give patience to women.
> Women give fullness to mountains.
> Celebrate each mountain, each woman.
> Sing songs to mountains and to women.
> Dance for them in your dreams.
>
> The spirit of mountains and of women
> Will give courage to our children
> Long after we are gone."

This book, *Spirit Walker,* is a very beautiful tribute to the powerful spiritual faith of Native Americans with its paintings by Frank Howell (whose work is widely exhibited) and Nancy Woods' poetry. Her book, *War Cry on a Prayer Feather* was nominated for a Pulitzer Prize in music in 1977.[69]

From Dr. Villoldo again: "Throughout time, human beings have looked upward for answers. This is why so many of our mythologies bring us to high places. The Shinto priests of Japan scale Mount Fuji; Turkey's Mount Ararat is revered by the Kurds; in India, Mount Arunachala is considered the embodiment of the Hindu god Shiva, the . . . Hopi kachinas revere Mount Blanca in Colorado; and, of course, the Greek gods lived . . . on Mount Olympus." We all know the story of Moses and Mount Sinai.[70]

Painting, sculpture, photographic art, and crafts are other routes to your inner creative Self. Some people can 'lose themselves' in their artistic creations. This is the frame of mind, 'the zone' you are looking for. I love to visit art galleries and so admire this wonderful gift. Beauty and Truth are synonymous; both help us find our 'real' selves. For a lot of us, music, whether creating or listening, can bring us to the Now moment better than anything else. There was a time when 'rockin' out' to Credence Clearwater could put me in that space and occasionally I can still enjoy that rhythm. As we evolve (and age) our 'loves' evolve as well. I probably developed or 'found' my love of New Age music from spending so much time on a massage table with a back problem. The music compositions of Steven

Halpern, pioneer in the field, have given me hours of relaxation. Reggae, world, jazz, and some classical music are also found in my collection. I listen to Hemi-Sync® music when on the treadmill; it just seems to 'fit.' Others like to have 'peppy' music when walking or running. Hemi-Sync® and Metamusic® were developed at The Monroe Institute, founded by Robert A. Monroe, from years of research into various aspects of human consciousness, from sleep research to psychic phenomena. They found that specific sound patterns could lead the brain to various states of consciousness, and create a resonance that results in focused, whole-brain coherence or 'hemispheric synchronization.' I especially like their Artists Series Vol. 1, "Inner Journey" and "Sleeping Through the Rain."[71] There is that word 'coherence' again.

Who can forget the passionate phrases from the soundtrack of Neil Diamond's *"Jonathan Livingston Seagull"*: "Be, as a page that aches for a word which speaks on a theme that is timeless," and "Lost on a painted sky," or "Sing, as song in search of a voice that is silent."[72] This, of course, is from the movie by the same name and the book by Richard Bach, author also of *Illusions: Adventures of a Reluctant Messiah* (1977). Recordings of Himalayan singing bowls, the sounds of the ocean, thunder and rain or other nature sounds can be very relaxing. I sometimes play *Pachelbel's Canon in D Major* or the *Ave Maria* until others in the room are pleading for something different!

One of my favorite artists is R. Carlos Nakai, playing, mostly, the Native American cedar flute. Nakai is a composer/performer of Navajo-Ute Heritage, and has a Masters degree from the University of Arizona and an honorary doctorate from Northern Arizona University. He has performed alone, in duets, quartets and with symphonies. His work ranges from traditional Native American ceremony music to jazz to Eastern and just about everything in between. From the cover of his CD, "Journeys": "At birth we embark on a good journey,* seeking a destination of happiness. The journeys on our life-road facilitate development of our emotional, mental, physical and spiritual states-of-being into a way of true power and wisdom. The Heart-center power, expressed as happiness and love, will guide us upward on a path away from frustration, bitter toil and travail. These journeys are directed inward, not, outwardly in material mementos of ego and possession. The lesson is relearning that which has been suppressed and forgotten in ourselves, since our earliest childhood."[73] You can see why his music is powerful. He has collaborated with Peter Kater, Nawang Khechog, Paul Horn, William Eaton and many others.

Only you know if music is your path to happiness or just a great addition to the journey. We've talked about Jonathan Goldman's work with healing sound; he put out a meditation CD called "Celestial Reiki II" with Laraaji and Sarah Benson.[74] I recently discovered the compositions of Deuter and love the tempo of his CD *"Atmospheres."*[75] The first track, "Uno," is very conducive to deep slow breathing. Deuter's *"East of the Full Moon"*[76] will take you to a very special place inside yourself.

Do you have a composition that you listen to over and over? Mine is "Quiet Heart/Spirit Wind" by Richard Warner.[77] The bamboo and alto flutes take me to my mountain, Mt. Wilson. The first time I heard the composition, my mind immediately went there and has ever since. Sometimes even our sweetest music, though, seems harsh when compared to the song composed by the wind in the pines – soothing, yet subtly exciting in its soft crescendos. Aldous Huxley said, "After silence, that which comes nearest to expressing the inexpressible is music."

Or is your passion for gardening or for the Earth? I know I love this big, blue planet! I have a photo taken over the western hemisphere from outer space by NASA in my Study and I adore it. The question of ecology and the environment has captured the attention of much of the world, but the big problems seem not to be improving. Lipton and Bhaerman wrote in the Preface, "*Spontaneous Evolution* introduces the notion that a miraculous healing awaits this planet once we accept our new responsibility to collectively tend the Garden rather than fight over the turf*." They further suggest, "Treating the land as landfill and our air, water, and soil as final resting places for pollutants is suicidal. Warfare, as a method of problem solving, has actually taken us to the brink of the ultimate solution to the human problem: no humans, no problem"[78] Strongly, but well said!

Did the Mayans have their paths to an inner connection with the Creator? Was it through Mother Earth, Nature, the Sun, studying the stars, making pottery, working the earth or sculpting in stone? Perhaps they were more like us than we have assumed.

Global Issues

Can environmental and political activism lead us to higher consciousness and greater global coherence? Some teachers versed in consciousness believe that the more one pushes against something and focuses upon it, the stronger it becomes and actually increases in your personal experience. Subjects, such as activism and politics, are difficult for me because, as you can see, I tend to look to the metaphysical, the mystic, the creative, the spiritual and the indigenous for inspiration and guidance. But we are all different and diversity is wonderful. As my dad used to say, "It takes all kinds to make up the world."

But we cannot finish our journey of discovery without looking at the reality of the problems our world is facing. We discussed some of the scientific findings in the fields of physics, astronomy, and even consciousness. In comparing those to the wisdom and predictions of the indigenous peoples, we found that in some ways, their lives may be or have been more successful than ours on some levels. An example of this comes from the Kogi people of Columbia, as Patricia Mercier relates from a woman of this culture, "In (2004) a special piece of low-lying land called La Luna that had been given to the Kogi, was designated an Indigenous Reserve, a protected area within one of UNESCO's Biosphere Reserves. Fifteen days later, without warning . . . a plane from Dyncorp passed over to fumigate with chemicals, supposedly to eradicate coca-growing." "Following this spraying with the illegal toxic mix Agent Green, La Luna became a tragedy. The Kogis had worked five years to regenerate the soil. Now they will have to wait at least five more years to replant there. Everything is contaminated and defoliated." "The destruction was pointless because there was no illegal coca growing in La Luna," her friend related.

Ms. Mercier continues, "The Kogi Elder Brothers and the Hopi together say that the imbalance in nature could cause the world to end. They do not understand why those in power fail to honour ancient Earth wisdom." She relates "an old Maya proverb: *Whoever cuts the trees as he pleases, cuts short his own life.*" And shares a Kogi prayer: (English translation)

"Only one thought
Only one Mother
Only one single word reaches upwards
Only one single trail leads heavenward."

She shares as well a quote from Tenzin Gyatso, the Fourteenth Dalai Lama "If we unbalance nature, humankind will suffer."[79] As Kahlil Gibran wrote in *The Prophet*,

"Forget not that the earth delights to feel your bare feet and the winds long to play with your hair."[80]

Author, John L. Peterson, founder of The Arlington Institute, a nonprofit future-oriented think tank, has the background to evaluate all the parameters of the global 'mess' facing us today. He was a Naval flight officer in both the Vietnam and Persian Gulf wars; he has worked with the National War College, the Institute for National Security Studies, the office of the Secretary of Defense and the National Security Council. His vision is practical, yet surprisingly consciousness-oriented. His book, *A Vision for 2012*, is small in size but 'jam-packed' with information and hope. He writes, "The coming years will strain the capabilities and emotional capacities of individuals, organizations, agencies and administrations in ways both strange and overpowering." The sources, from conventional science, business and economy, ecological and climate studies, to the unease of the common people and indigenous wisdom all point to extraordinary years around 2012. He says. ". . . analysts suggest that the period between 2005 and 2015 will see a confluence of economic, social, cultural, ecological, technological, geopolitical, and military distress and disruption of unprecedented magnitude." "(A) recent BBC headline (reads): 'Current global consumption levels could result in a large-scale ecosystem collapse by the middle of the century, environmental group WWF has warned'." "*The Washington Post* wrote, 'Birds, bees, bats and other species that pollinate North American plant life are losing population, according to a study released yesterday by the National Research Council.'"

Speaking about the population issue in developing countries, Peterson writes, "In addition to the social stability issues, these millions of people are most at risk from threats like global pandemics and would produce an overwhelming governmental and social system failure in the face of such an event." On the global oil situation he predicts, "When the fact that available oil has peaked starts to become generally known, a global competition for the dwindling supply will ensue." ". . . if China is confronted with the choice between domestic instability or using violence to secure access to decreasing supplies held by another country, will that be a hard decision?"[81]

Then there is the issue of climate change, for which there are some disheartening discoveries. The melting of polar ice "means thawing permafrost, which releases large amounts of methane (ten times more effective than CO2 at producing greenhouse gases) into the atmosphere," Peterson reveals. "Glaciers in the Himalayas are receding faster than in any other part of the world and, if the present rate of retreat continues, they may be gone by 2035. More than 2 billion people – a third of the world's population – rely on the Himalayas for water." Peterson quotes writer Tara Lohan, "In the U.S., there are currently 150 new coal-fired power plants on the drawing board. The amount of polluting emissions they will release is staggering . . ." "This alone will basically negate every other effort currently being considered to fight climate change."

In 2004 a former commerce secretary gave figures about the national budget that predicts by 2020 Social Security and Medicare will have an unfunded annual deficit of $783 billion.[82] The news is not good, but then Peterson previews what might happen if a global epidemic of, say bird flu, should occur today. "Computer estimates built around the propagation rate of previous epidemics suggest that this one, if it comes, will be much faster and more deadly than the last major outbreak in 1918, which killed 50 to 100 million people over a period of twelve months. Projections vary from 300 million to over 1 billion people . . . and would probably run the world in about six months." "It would be a very quick and deadly blow to the world." This is only one of several possible scenarios confronting us. In light of recent terrorism, there is the threat of nuclear or biological warfare; then, due to climate changes, there is the increased threat of bigger hurricanes, tornadoes, earthquakes and volcanic eruptions, such as the supervolcano that lies under Yellowstone National Park and the La Palma volcano in the Canary Islands that could cause a huge tsunami. The five big stresses, as we know, are *energy, economy, demographic* (differentials in population growth rates between rich and poor societies and expansion of megacities in poor societies), *environmental*, and *climate stress*.[83]

There are also, according to Peterson, many developments worldwide to give us HOPE. "We are becoming ever more closely connected in increasingly interdependent ways." "We know we're closely connected to a rapidly changing world that is headed . . . in a new direction – but it's not clear where it's gong." "This is a unique time in history; something big is in the works." "Humans don't deal well with discontinuities and rapid change. We build our perspectives and options around the past, not generally informed about potential futures." In regard to this statement, I like the famous quote by Will Rogers, "Quit lookin' back. We ain't goin' thatta way!" Peterson continues, "Scientists have estimated that technology and knowledge is exploding a million times faster than the rate at which our underlying social and cognitive frameworks change." "The future doesn't just happen; we make it happen. It is the product of our desires, interests, perspectives, visions, and actions. What we think and what we do makes a difference. It makes the only difference."* Dr. Aubrey de Grey, Cambridge University researcher, "believes that if you . . . are about sixty years old, you may well live to be two hundred. A series of advances will show up, one after another, just in time. . ."[84]

Corinne McLaughlin, author of *The Builders of the Dawn*, asks, "Why not imagine a better world in the future and live this vision moment to moment in the choices you make and the work that you do? In this way, you can help create it." "Did you know that at this very moment a new world is emerging right through the cracks and crevices of the old world?" "By 2012, this new world, born out of the creative minds and compassionate hearts of self-empowered visionaries everywhere, will be even more visible and influential, affecting every aspect of life." "It's easy to see the negative impact of business and politics in the world today. . ." But McLaughlin sees evidence of improvement: "Lifestyles of Health and Sustainability (LOHAS) reports there is now a $228.9 billion U.S. marketplace for goods and services focused on holistic health, the environment, social justice, personal development, and sustainable living." *Business Week* reported that 95% of Americans reject the idea that a corporation's only purpose is to make money, and 39% of U.S. investors say they frequently check on business practices, values, and ethics before investing." "In recent years, over 300 multinational corporations have joined the United Nations Global Compact, pledging to support environmental protection, human rights, and higher labor standards. Companies not reporting significant progress in these directions each year are eliminated from the compact," she reports in her chapter in *The Mystery of 2012. (Sounds True)* Whole Foods, the only Fortune 500 company to offset 100 percent of its electricity use in all 180 stores, is purchasing more than 458,000 megawatt-hours of renewable energy credits from wind farms.[85]

Ms. McLaughlin records some progress in politics; she "coordinated a national task force on sustainable communities for President Clinton's Council on Sustainable Development, which brought together adversaries – corporate and environmental leaders – to find common ground and build a consensus on environmental protection and economic development." She cites the nonviolent change in the South African government and the transition from apartheid. "A growing spiritual activism of the progressive left in the U.S. is countering a decade of intense activism by the fundamentalist right." "Positions may appear mutually exclusive, while interests tend to overlap, and this is the key to having all sides work together to transform conflict. SFCG (Search for Common Ground) calls this 'cross-stitching,' a way to reweave the whole." (There's that tapestry again!) This organization has been successful in organizing high-level meetings (between such groups as Iranians and Americans), and helping opponents in the abortion debate find that their common interest was preventing unwanted pregnancies and making adoption more easily available. They have pulled together dialogues between Arabs and Israelis, environmentalists and developers, and Republicans and Democrats. (Remember our chapter on Non-Duality?)

"In 2006, a study by the prestigious Princeton Survey Associates found that a remarkable 85% of Americans agree that the country 'has become so polarized between Democrats and Republicans that Washington can't seem to make progress solving the nations' problems'." "One of the new political approaches . . . is a synthesis of left and right . . . the *best* in both liberal and conservative perspectives is highlighted on an issue to create a higher synthesis and more innovative policies." McLaughlin quotes from President Barack Obama's book, *The Audacity of Hope*, "The next generation is to some degree liberated from what I call the either/or arguments around these issues." She concludes, "All of these growing trends . . . give me a real sense of hope for the future. . ." "The choice is up to each of us."[86] Carlos Barrios, from whom we heard earlier, is quoted as saying: "Some observers say this alignment with the heart of the galaxy in 2012 will open a channel for cosmic energy to flow through the earth, cleansing it and all that dwells upon it, raising all to a higher level of vibration.[87]

When *The Audacity of Hope* was published Barack Obama was a Senator from Illinois; of course, today he is the President of the United States. I bought the book because I liked the title, since I've been 'preaching' hope for a long time and found a lot of people thought that I was out of touch with reality. I have liked Barack Obama ever since the first speech I heard him make in his bid for the highest office of our country. I like many of the ideals he expresses in the book, such as, "I am angry about policies that consistently favor the wealthy and powerful over average Americans . . ." "I believe in evolution, scientific inquiry, and global warming; I believe in free speech, whether politically correct or politically incorrect, and I am suspicious of using government to impose anybody's religious beliefs – including my own – on nonbelievers." ". . . I can't help but view the American experience through the lens of a black man of mixed heritage. . ." "I believe in free market, competition, and entrepreneurship, and think no small number of government programs don't work as advertised. I wish the country had fewer lawyers and more engineers." ". . . I carry few illusions about our enemies, and revere the courage and competence of our military. I reject a politics that is based solely on racial identity, gender identity, sexual orientation, or victimhood generally."[88]

I learned several things from this book, which is not surprising, since politics has never been my 'strong suit' or much of an interest for that matter. This is the first President since John F. Kennedy that speaks his ideals clearly enough to be of interest to me. (There I go, dating myself again. I was a newlywed when JFK was assassinated and I will never forget that day.) I learned: "The Constitution makes no mention of the filibuster; it is a Senate rule, one that dates back to the very first Congress." "The only way to break a filibuster is for three-fifths of the Senate to invoke something called cloture – that is, the cessation of debate." He relates how filibuster was used frequently during the civil rights era to "choke off any and every piece of civil rights legislation before the Senate. . ." "For many blacks in the South, the filibuster had snuffed out hope."[89]

I especially liked Obama's recounting of his discussion with Warren Buffet (second richest man in the world at that time) about taxes. Buffet "wanted to know why Washington continued to cut taxes for people in his income bracket when the country was broke." He said he had never used tax shelters, but "I'll pay a lower effective tax rate this year, than my receptionist." Obama continues, "Buffet's low rates were a consequence of the fact that, like most wealthy Americans, almost all his income came from dividends and capital gains, investment income that since 2003 (had) been taxed at only 15%. The receptionist's salary . . . was taxed at almost twice that rate once FICA was included. From Buffet's perspective, the discrepancy was unconscionable." Buffet concluded, "And it just makes sense that those of us who've benefited the most from the market should pay a bigger share." Obama relates, "We spent the next hour talking about globalization, executive compensation, the worsening trade deficit, and the national debt. He was especially exercised over Bush's proposed elimination of the estate tax, a step he believed would encourage an aristocracy of wealth rather than merit."[90] As of this date, some of these issues are still being debated.

When President Obama received the Nobel Peace Prize, the International Campaign for Tibet newsletter quoted from their blog, writing, "Obama sounded most like His Holiness (the Dalai Lama) when he said, ". . . we do not have to think that human nature is perfect for us to still believe that the human condition can be perfected. We do not have to live in an idealized world to still reach for those ideals that will make it a better place. The nonviolence practiced by men like Gandhi and (Martin Luther, Jr.) King may not have been practical or possible in every circumstance, but the love that they preached – their faith in human progress – must always be the North Star that guides us on our journey."* This kind of ideals gets my vote.

I am as big a fan of democracy as anyone, but does it seem to you that they have a tendency to become 'top-heavy?" For example, as more Representatives and Senators are elected, the outgoing ones continue to draw pensions, health insurance benefits, etc. I am curious how many people are now in that category, on the payroll. But as Mike Dooley pointed out: when we speak of 'the government' as though it were some conspiring entity 'up there' somewhere we forget that in a democracy 'we are the government.' We have a voice; we are not powerless. In his DVD "A Wrinkle in Time," he points out that many who complain we should fund a healthcare system are the ones who choose, instead of health insurance, to have jet skis, a big-screen TV, high-speed Internet access, etc. This is rather 'tongue in cheek,' but actually very true in many cases. The problem seems to be a failure to make the hard choices and accept responsibility for how we are creating our future.[91]

We may look at the Mayans, and more so, the Toltecs, and find their practice of human sacrifice as barbaric, but let's look at our world today:

- A fascination with horror, war, and 'grisly' movies and TV brings in millions of dollars for the entertainment industry.

- War, killing, maiming, even torture continues.

- The most-watched TV shows include scenes of crime, rape, autopsy, bodies decaying, people injured or killed.

- Video games featuring destruction, murder, annihilation are promoted for children.

- We routinely read, hear or see incidents of pornography, perversion, and molestation.

- Our prisons are full and crime takes a huge amount of our national and state budgets.

Before we become self-righteous perhaps we should look at our own culture. It is obvious there are some 'problems' in our world, but, as we saw in the chapter on Consciousness, how we direct our thoughts and what we believe make a difference in how we approach those issues. Thoughts, beliefs, intentions are three very powerful words in creating our future. John Peterson mentions 'wild cards' in *A Vision for 2012* calling them "low probability, high impact events that are so big or arrive so fast that underlying social systems cannot effectively deal with them. A global bird flu pandemic is a potential wild card. . ." Other possibilities that he mentions are: a close call with an asteroid or comet, bacteria becoming immune to antibiotics, or the U.S. Federal Reserve fails; or on the positive side: altruism breaks out, time travel becomes possible, or the return of the Awaited One.[92] In the case of the negative 'wild cards,' how one perceives the event will make a huge difference. The positive ones? Be thankful! Peterson concludes, "We are all blessed to live at this period of extraordinary transformation. Each in his or her own way has a special role to play in contributing to the ultimate shape and function of this new world; that is probably why we are here at this time." He notes some websites that give information, suggestions, ideas and connections for working together to cocreate a positive future.[93] As we have heard before from Albert Einstein, "Problems cannot be solved at the same level of consciousness that made the problems." Our collective feelings can be a self-fulfilling prophecy, for evolving or suffering.

Another website that will give you yet another perspective is that of Kiesha Crowther, a Wisdom Keeper, called the 'Little Grandmother.' She writes there, "We are the Tribe of Many Colors – the Rainbows, prophesied to come during the great time of change when the world would transform both physically and spiritually, a time that will bring great enlightenment." Jamie Sams, author and artist of the Seneca and Cherokee Nations supports Crowther by writing, "These Two-legged will be called the Rainbow Tribe, for they are the product of thousands of years of melding among the five original races. These Children of Earth have been called together to open their hearts and to move beyond the barriers of disconnection. The medicine they carry is the Whirling Rainbow of Peace, which will mark the union of the five races as ONE."[94] Kiesha is a Native American shaman and "her primary responsibility is to be a shaman for the Tribe of Many Colors – which includes non-indigenous people of all backgrounds." She has been recognized as such by the spiritual elders of many indigenous tribes and peoples including the Cherokee, the Cheyenne, the Hopi, the Inuit, the Aboriginal people, the Waitaha, the Maori, the Maya, and the Zulu, as well as the lamas of Nepal and Tibet." She is "being guided by elders who comprise the Continental Council of Indigenous Elders from indigenous peoples all over the planet, as well as receiving direct teaching from spirit guides and indigenous elders past of several different traditions, including the Sioux and Salish." "She does not teach the spiritual ways and customs of any particular tribe, but the wisdom she is taught by her spirit guides . . . that crosses cultural boundaries." ". . . she has been taught that the time has come for all people to unite as children of one Mother, and to recognize a common brotherhood within the heart's wisdom – that is neither indigenous nor non-indigenous. . ."

As I said, many paths, and we must allow each person to have and follow their own kind of path which may not resemble yours or mine at all. Were the shamans of the Maya and other indigenous peoples able to 'see' the 'karma' of future mass consciousness? Even we can almost see when a certain attitude or action will bring a certain result. Could they foretell when a planetary/cosmos alignment coupled with an attitude/action of the majority would result in an 'event?' Probably.

John Lennon wrote the lyrics to a very wonderful song and it is quoted on Poetic History page of the Global Consciousness Project website.[95] GCP has logistical support from the Institute of Noetic Sciences, founded by Edgar Mitchell, astronaut on the Apollo 14 mission.

> Imagine
>
> Imagine there's no heaven
> It's easy if you try
> No hell below us
> Above us only sky
> Imagine all the people
>
> Imagine there's no countries
> It isn't hard to do
> Nothing to kill or die for
> And no religion too

Imagine all the people
Living life in peace

Imagine no possessions
I wonder if you can
No need for greed or hunger
A brotherhood of man
Imagine all the people
Sharing all the world

You may say I'm a dreamer
But I'm not the only one
I hope someday you'll join us
And the world will live as one

- John Lennon

There is a quote from *The Course in Miracles* in one of Wayne Dyer's books, *There's a Spiritual Solution to Every Problem,* which might help in many difficult situations: "I can choose peace, rather than this." In Part II of this book he uses The Prayer of St. Francis as a framework for "Putting Spiritual Problem Solving into Action. Have you read that prayer lately? Some of its main points are his chapter headings:

- Lord, Make Me an Instrument of Thy Peace

- Where There is Hatred, Let Me Sow Love

- Where There is Injury, Pardon

- Where There is Despair, Hope

- Where There is Sadness, Joy [96]

When we through joy feel ourselves lifted, rising higher, we see there are unlimited octaves of beauty and happiness that we can experience.

The Perfection of Hope, Harmony and Love

The Mayans believed there would be great changes at the end of the Fourth Sun, but that the Fifth Sun would be, as Lawrence Joseph learned from modern Mayans, "the best shot in the past 26,000 years for humanity to become enlightened and move closer to the gods."[97] It appears that there are 'wild cards' in the subject of consciousness as well. From our chapter on Consciousness, we saw that by directing our thoughts and intentions we are capable of creating and responding to life in a much more positive way. But to the wild cards: As I was doing my back exercises a few days ago, my eyes fell on a book and I felt an urge to take it down. I had bought it several years ago, but never read all of it. It gave me a final clue to some of the issues we have been examining.

It was the final, ninth, book in Neale Donald Walsch's *Conversations With God* series, *Home With God*. I had not really understood why even with our best efforts, life just seemed to take a different route at times. This book (God, The All That Is) explains the subconscious, conscious and superconscious in ways we can all understand. We create from all three of these levels. This is what I call a 'wild card,' because we may be totally unaware that we are creating our lives from all these levels. "The subconscious is the place of experience which you do not know about, or (from which you do not realize you) consciously* create your reality." "This is not a 'bad' level of experience, so do not judge it. It is a gift, because it allows you to do things automatically . . . like growing your hair, blinking your eyes, or beating your heart. . ." "The subconscious also creates instant solutions to problems . . . automated response based on prior data . . . It can save your life. Yet if you are unaware of what parts of your life you have chosen to create automatically, you could imagine yourself to be at the 'effect' of life rather than at cause in the matter. You could even create yourself as a victim." At the conscious level you create with some awareness. "How much . . . you are aware depends upon your 'level of consciousness.'" The Superconsciousness Level is the place of experience at which you . . . create your reality with full awareness . . ." This is the soul level. Most of you are not aware at a conscious level of your superconscious intentions . . ." This level "holds the larger agenda of the soul – which is to move to Completion in what you came to the body to experience. . ." (It is) "constantly leading you to your next most desired growth experience, drawing to you the exact, right, and perfect people places, and events with which to have that . . . creating Awareness of your True Being."

This now gives us more insight into the prayer, "God, Thy will be done." Perhaps we should ask, "May I be more aware of my Higher Intentions." Meditation helps us 'pull in' those levels of consciousness we are usually not aware of. Walsch asked about making our intentions the same on all levels and God answers, "This three-in-one level . . . is called the *supra*consciousness." "Some people (go there) in meditation, others in deep prayer, others through ritual or dance or through sacred ceremony. . ." "All three levels of consciousness have become one." "Supraconsciousness is not simply a combining" but also a transcending. "Beingness is the Ultimate Source of Creation within you."[98] To me the perfection is found in allowing and Allowing helps us find the perfection.

There are eighteen Remembrances in *Home With God*; most are explaining our journey Home. The Fourth reminds us that "No path back Home is better than any other path." He also writes, "There are no victims and there are no villians." (You may want to read the book to fully understand that!) But he goes on to say, you may think you are a victim and *"you will experience yourself as a victim* in spite of the fact that you are not." "It is impossible to be a victim of circumstances you create." Or you may see others as victims, "in spite of the fact that he or she is not." "You may experience whatever you choose . . . you have free will." "If it is true that God is the Creator, this means that you, too, are a creator. God creates *all* of life, and you create all of *your* life." "Perspective creates perception, and perception creates experience. The experience that perception creates for you is what you call (your) 'truth'." "You can decide what you want to see, and then, having placed it there, you will find it there."[99] It is a matter of choice; as The Teachings of Abraham® say, "You are so free, you can choose bondage."[100] But as Walsch writes in the Introduction to this book, "The destination is the same for all of us. We are all on a journey* Home, and we shall not fail to arrive there."

In *Return to the Sacred*, Jonathan H. Ellerby includes one of his poems:

> I see through the ancient archways
> the sacred door is open
> in the distance a traveler comes
> through time
> and from the heart of spirit
> I know the journey well,
> I watch the moving form, though
> I cannot tell if they are coming or going
> is the sun rising or setting
> is it man or woman
> the path shines like a thin deep river
> shadows and light play tricks on the mind
> but none of this matters
> this is the arrival and the departure
> there is only one direction:
> inward
> the road is littered with scrolls and beads
> talismans, scriptures, rattles, and robes
> every step is sacred
> and none of it matters
> for the road only leads one place:
> home.
>
> -- J. H. Ellerby[101]

Since hope is one of my reasons for writing this book, I must include what the Creator (through Neale Donald Walsch) says about this subject in *Home With God*, "Hope plays a wonderful role in 'death' and in 'life'. (They are the same, of course.) Never give up hope. Never. Hope is a statement of your highest desire." "Hope is thought, made Divine." "Hope is the doorway to belief, belief is the doorway to knowing, knowing is the doorway to creation, and creation is the doorway to experience. Experience is the doorway to expression, expression is the doorway to becoming, becoming is the activity of all Life. . ."[102]

Perhaps fear of death is one of our strongest obstacles on the journey, but both Walsch and Abraham-Hicks seek to allay our concerns. "The Seventh Remembrance (Walsch) states: Death does not exist," and that is explained fully.[103] Esther/Abraham calls the death experience 'croaking,' as they say, to be as disrespectful as possible, because *there is not death*. "There is no end to life experience, for you are an Eternal Being with never-ending opportunities for joyous expansion."[104] Perhaps the word 'death' will disappear from our vocabulary. Or as Abraham says, "You can't get it wrong, and you can't get it (all) done." There is a pervasive core of Well-Being.[105] We are always 'in process.' So enjoy the journey. It is fail-safe!

We saw how the connectedness of the physical and nonphysical worlds is becoming more apparent. The orbs showing up in digital photography, as well as other phenomena that lift that 'veil' between dimensions, seem to be trying to show us the unity of worlds seen and unseen. The Heinemann's (the authors of *Orbs*) remind us that "the circle or sphere . . . represents oneness, wholeness, unity . . . coming together as one." "Perhaps their (orbs') visibility is intended to wake us up to . . . 'All is one,' 'We are one,' 'One earth, one humanity, one spirit . . . one people, interconnected, interrelated, with one destiny." "This seems to tell us that we are not alone but are assisted and guided . . . that there is a field of infinite possibilities to tap into."[106] Patricio Domínguez, a Native American spiritual advisor, revealed the following to authors Chris Morton and Ceri Louise Thomas: ". . . in the Native American world view . . . everything is sacred . . . both living material and inorganic material. Everything that exists . . . is a part of the whole, a part of the sacred essence of all existence." "Our physical existence is only one level of the greater reality, which has many layers, many dimensions, and many paths through which everything is connected and through which everything is . . . part of the one great whole." "Humanity has a choice about which direction to take."

"We can take a new path by changing our awareness of our role here on Mother Earth."[107] Everything around us – the air, trees, animals, rocks – is in harmony with Source; only we humans can restrict the flow of all this harmony, good, abundance and Well-Being. This is a very powerful alliance when you 'line up' with all the Well-Being that is in nature, trees, the animals, the planet, and the universe.

Eckhart Tolle reveals, "There are three words that convey the secret of the art of living . . . One With Life. Being one with life is being one with Now."[108] Drunvalo Melchizadek shares that his "stories are given . . . so that you also will remember and return into the harmony and flow of the Universe. Love is the answer to every question – even the questions of the mind."[109] Revealing the greatest force in our universe – that of Love, Rhonda Byrne, in *The Power,* quotes Charles Haanel, a New Thought author, "The law of attraction or the law of love . . . they are one and the same." "The measure of love is love without measure." Saint Bernard of Clairvaux (1090-1153 AD) - Christian monk and mystic, said that.

Mildred Lisette Norman (1908-1981) who took the name Peace Pilgrim said, "Love is the greatest power on earth. It conquers all things." Mother Teresa (1910-1997) believed, "It's not how much we give but how much love we put into giving." Is it any wonder she was awarded the Nobel Peace Prize. Rhonda Byrne has brought together many great thoughts about Love in her book *The Power,* stating: *"Love is the Power that Connects Everything."* She quotes the Chinese philosopher, Mozi (circa 470- 391 BCE), "When all the people in the world love one another, then the strong will not overpower the weak, the many will not oppress the few, the wealthy will not mock the poor, the honored will not disdain the humble, and the cunning will not deceive the simple." Ms. Byrne suggests, "To find heaven on earth is to live your life at the same frequency as your being – pure love and joy." The formula for the Creative Process for her is: "Imagine it. Feel it. Receive it."[110]

Albert Einstein once said, "There are only two ways to live your life. One is as though nothing is a miracle; the other is as though everything is a miracle. I prefer the second. The way to begin creating some of those miracles in your life is with unconditional love. No one can show us that path as clearly as Marci Shimoff, featured in *The Secret,* who wrote *Happy for No Reason* (among other books). *Love for No Reason* (with Carol

Kline) is the clearest, most complete, examination of Love you can read; it includes conversations with over a hundred 'real life' people as examples and there are exercises to help you get there. It is focused around three themes: (1) "Love is who we are." "Love is more like an ocean that's inside and all around us." "Experiencing that you *are* love is the ultimate form of self-love. (2) "The purpose of life is to expand in love." ". . . all the accounts I've read about near-death experiences point to the importance of focusing on love." ". . . at the end of our lives . . . our souls are asked, 'How much did you love?'" (3) "The heart is the portal to Love." ". . . keeping your (spiritual) heart open is the goal of all the practices, tools and techniques in the *Love For No Reason* program. . ."

These authors quote Emmet Fox at the beginning of the book:

> There is no difficulty that enough love will not conquer. . .
> No door that enough love will not open
> No gulf that enough love will not bridge,
> No wall that enough love will not throw down. . .
> It makes no difference how deeply seated may be the trouble,
> how great the mistake,
> sufficient realization of love will resolve it all.
> If only you could love enough,
> you would be the happiest and most powerful being in the universe.[111]

When I think about the perfection of our world, it is hard to wrap my mind around it. Scientists believe, based on the theory of general relativity, with the high energy density, huge temperatures and pressures, and the very rapid expansion and cooling after the Big Bang of the Singularity, after approximately 10^{-37} part of a second a 'phase transition' caused a cosmic inflation (which is Inflation with a huge capital I!) which led to the formation of the elemental particles such as hydrogen and helium. The interesting point about all this is, according to our leading edge scientists, had this process that occurred in the first few seconds NOT happened exactly the way it did, the universe as we know it could not have formed. Had hydrogen decay been a minute amount smaller or larger, there would be no physical universe. Talk about Perfection! Balance on such an enormous scale is awe-inspiring. The hydrogen formed in the first few minutes after the Big Bang some 13.7 billion years ago could be in the water you are drinking at this moment. But we only have to be aware of the trillions of cells and the thousands of processes going on every minute in our bodies to be amazed. And those are only two tiny examples of the magnitude of perfection in our physical world; there are probably dimensions we don't even know exist.

When I consider that I am an interface, a transmission line, so to speak, between this physical, earth, matter, time dimension and the spirit world, Sky Father as the Mayans said, and higher dimensions, I am humbled. The energy/life force flows into us, mixes and becomes manifest creation. And our thoughts are a huge factor in HOW and WHAT that creation becomes. The more I learn about the perfection of the universe, our Mother Earth, the cycles, the expanding nature of life and consciousness, the more I wonder about the perfection of the Unmanifest, Spirit world, or heaven, as some say, from which All flows out to us.

There is a beautiful little book written by Stephen C. Paul called *Illuminations: Visions for Change, Growth and Self-Acceptance,* that has a wonderful painting by Gary Max Collins with each of the aphorisms. It is hard to pick a favorite; they are all great, but these fit our discussion here. "The earth becomes heaven when you release your fear." "Allow the earth to be a part of heaven." And when we contemplate how to do that, it appears to come down to minute by minute choices. As this book encourages, "Everything and everyone in your life is there by your choice." I hope we all know by now that when you STOP giving your attention to something NOT wanted in your life, it will disappear. "To be free yourself, you have to release everyone else." "There are people who will joyfully greet the person you are becoming."[112]

The Mayans may have been concerned that we, all of us together, might not be able to 'handle' the stresses that might effect Mother Earth at the end of the Fourth Sun, but I believe we just need to get in touch with and enlarge the inherent Well-Being that is at the core of everything. As the Abraham materials relate, "The unlimited Consciousness of the particles that make up your immense planet never,* under any circumstances, worry themselves out of alignment with the Energy of their Source. The core Vibration of Well-Being is so significant that departure from it does not occur."[113] They also assure us on their Placemat Process writings, "We have been speaking of a time of awakening, and while you are not literally asleep, you have, for a very long time, been suppressing an important part of yourself. In the time that is before you, many of you will awaken that important part of your awareness, and perceive from a broader perspective."[114]

Barbara Marx Hubbard "reminds us that . . . *things happen the way they happen, when they happen, where they happen, for a reason – all of which is in perfect order, reflecting the perfection of the Universe and of Life itself.*" Echoing what we saw earlier, Neale Donald Walsch writes, "Life works in mysterious ways to produce the wonder, beauty, and *intricacy* of its tapestry.* One thread out of place, one weaving incomplete, and the entire picture has been altered. Indeed, it has been said that if one tiny chemical interaction out of a million had been different, the Universe as we know it would have been impossible."[115]

You are loved by All-That-Is because that state of loving Well-Being is 'built in;' it's flowing to all of Creation. This Spirit/Energy is Always flowing to us and when we allow these higher frequencies to flow on through us, it feels like joy, compassion or bliss; this is 'bringing heaven onto earth.' Source cannot force this on us. It is always available, but you and I must 'open the portal;' we have free will, we can be aware of that connection and allow its expression, or not. In moments of bliss, watching a sunset or playing with your dog, you can feel the expansiveness of your Being, as though you were boundless and unlimited. From this 'space' you can connect with any facet of our multidimensional world, and you are making the tapestry of your life more stunning and wonderful by the second. So, here is another word to delete: delete from your vocabulary: struggle. My 'job' here, yours too, is to 'seek joy,' 'follow my bliss,' look for ways to feel appreciation and love; the rest is just 'banging it out,' as they say.

James Twyman, author of ten books, has been called a 'Peace Troubadour'; he writes, the Creative Force, also known as God, says: "You are holy. You are perfect. You are safe."[116] Even though the evening news tells us about wars, injustice, tsunamis, floods and the list goes on and on, I want to assure you that each and every one of you are cared for and adored by All-That-Is, the total Well-Beingness of the Universe. You are so supported by the perfection of Life and the Afterlife that you CAN 'love it all.'

Your Inner Being, Higher Self, eternal 'part' knows this and is constantly nudging you to come back to the love and ease that are your true beingness or nature. Why are we here? For the Joy! "Laughter is the key to everything. It is far more powerful than prayer, than meditation. It is the stuff of which the world is created. Find laughter, find freedom."[117] (Whitley Strieber)

Wherever you are right now is just fine; this is your launching 'pad.' If you want something to be different, you are sending out that signal without even saying a word. The Universe hears you. Now, you can choose thoughts that make you feel good, thoughts about how YOU want your life to be. Love the world into change. Love every person into their highest manifestation of Being. As Abraham through Esther say, "Nothing is more important than your feeling good." Remember, you are perfect. You are always 'in process.' And your thoughts, words, and actions each minute are important to this time/space reality. We can each make the journey as fulfilling as the destination. For any question, problem or issue, the answer, the 'bottom line' is always:

Love

* indicates author's emphasis or explanation placed within a quote

ENDNOTES

INTRODUCTION

1. Geoff Stray, *Beyond 2012: Catastrophe or Awakening?* (Rochester, VT, Bear and Company, 2005, 2006, 2009)
2. www.renaissanceastrology.com/hermestrimegistus.html
3. Gregg Braden, *The Isaiah Effect: Decoding the Lost Science of Prayer and Prophecy*, (NY, NY, Harmony Books, Crown Publishing Group, Random House Inc., 2000)
4. Sounds True, Inc., *The Mystery of 2012: Predictions, Prophecies & Possibilities*, edited by Tami Simon, (Boulder, CO, Sounds True, Inc., 2007) chapter by Sharron Rose, "2012, Galactic Alignment, and the Great Goddess"
5. Immanuel Velikovsky, *Earth in Upheaval*, (New York: Doubleday & Company, 1955),(New York: Pocket Books, A Division of Simon & Shuster, Inc., 1977)
6. Gregg Braden, *The Divine Matrix*, (Carlsbad, CA, Hay House, Inc., 2007); *Fractal Time*, Carlsbad, Hay House Inc., 2009)
7. Peter James and Nick Thorpe, *Ancient Mysteries*, (New York, Toronto, The Ballantine Publishing Group, a division of Random House, Inc. and Random House of Canada Limited, 1999)
8. Graham Hancock, *Fingerprints of the Gods*, (Toronto, Ontario, Doubleday Canada Ltd., 1995) page 476.
9. www.Foxnews.com/scitech/2011/01/06/magnetic-north-pole -
10. Gregg Braden, *The Isaiah Effect*, (NY, NY, Harmony Books, 2000), quoting Tom Majeski, *St. Paul Pioneer Press*, Oct.7, 1997.

THE MAYANS

1. John Major Jenkins, *The 2012 Story: The Myths, Fallacies, and Truth Behind the Most Intriguing Date in History*, (NY,NY, Jeremy P. Tarcher/Penguin, 2009)
2. http://en.wikipedia.org/wiki/Clovis_culture

3. Ibid
4. http://en.wikipedia.org/wiki/Clovis_culture
5. http://archaeology.about.com/b/2008/04/28/clovis
6. Chris Morton and Ceri Louise Thomas, *The Mystery of the Crystal Skulls: Unlocking the Secrets of the Past, Present and Future*, (Rochester, VT, Bear & Company, 1997, 1998, 2002)
7. Adrian Gilbert and Maurice Cotterell, *The Mayan Prophecies: Unlocking the Secrets of a Lost Civilization*, (Shaftesbury, Dorset, UK, Element Books Ltd.,1995)
8. Zechariah Sitchen, *Journeys to the Mythical Past*, (Rochester, VT, Bear & Company, 2007)
9. Graham Hancock, *Fingerprints of the Gods*, (Toronto, Ontario, Doubleday Canada Ltd., 1995) pg. 104.
10. Ibid. Citing Peter Tomkins, *Mysteries of the Mexican Pyramids*, (London, Thames & Hudson, 1987.)
11. Ibid
12. Carlos Barrios, *The Book of Destiny: Unlocking the Secrets of the Ancient Mayans and the Prophecy of 2012*, (New York, NY, HarperCollins Publishers, 2009)
13. http://www.hindunet.org/vedas/rigveda/
14. Sounds True, Inc., *The Mystery of 2012: Predictions, Prophecies & Possibilities*, edited by Tami Simon, (Boulder, CO, Sounds True, Inc., 2007). "2012 and the Maya World" by Robert K. Sitler, page 89.
15. Barbara Hand Clow, *Catastrophobia: The Truth Behind Earth Changes in the Coming Age of Light*, (Rochester, VT, Bear & Company, 2001)
16. Ibid
17. Carlos Barrios, *The Book of Destiny: Unlocking the Secrets of the Ancient Mayans and the Prophecy of 2012*, (New York, NY, HarperCollins Publishers, 2009)
18. http://en.wikipedia.org/wiki/Great_Pyramid_of_Cholula
19. Graham Hancock, *Fingerprints of the Gods*, (Toronto, Ontario, Doubleday Canada Ltd.,1995)
20. David Freidel, Linda Schele, & Joy Parker, *Maya Cosmos: Three Thousand Years on the Shaman's Path*, (New York, NY, William Morrow and Company, Inc., 1995)
21. Ibid
22. http://ascendingpassage.com/plato-atlantis-critias.htm
23. Ralph Ellis, *Thoth: Architect of the Universe* (Dorset, England, Edfu Books, 1997), 205. Cited by Barbara Hand Clow, *Catastrophobia: The Truth Behind Earth Changes in the Coming Age of Light*, (Rochester, VT, Bear and Company, 2001)

24. Whitley Strieber, *The Key: A True Encounter,* (Copyright 2001, 2011 by Walker & Collier, Inc., Published by Jeremy P. Tarcher/Penguin, New York, 2011)
25. Graham Hancock, *Fingerprints of the Gods,* (Toronto, Ontario, Doubleday Canada Ltd., 1995), quoting Peter Tomkins, *Secrets of the Great Pyramid,* (New York, Harper & Rowe, 1978)
26. Andrew Collins, *Gateway to Atlantis: The Search for the Source of a Lost Civilization,* (London, Headline Book Publishing, 2000) from Geoff Stray, *Beyond 2012: Catastrophe or Awakening?* (Rochester, VT, Bear and Company, 2005, 2006, 2009)
27. Ibid
28. Immanuel Velikovsky, *Earth in Upheaval* (New York, Pocket Books, Simon & Schuster, 1977)
29. Geoff Stray, *Beyond 2012: Catastrophe or Awakening?* (Rochester, VT, Bear and Company, 2005, 2006, 2009) citing Andrew Collins, *Gateway to Atlantis: The Search for the Source of a Lost Civilization,* (London, Headline Book Publishing, 2000)
30. Immanuel Velikovsky, *Earth in Upheaval* (New York, Pocket Books, Simon & Schuster, 1977)
31. D.S. Allan and J.B. Delair, *Cataclysm! Compelling Evidence of a Cosmic Catastrophe in 9500 B.C.,* (Sante Fe, Bear & Company, 1997)
32. Barbara Hand Clow, *Catastrophobia:The Truth Behind Earth Changes in the Coming Age of Light,* (Rochester, VT, Bear & Company, 2001)
33. Ibid
34. http://etheric.com/Galactic Center/GRB.html
35. Graham Hancock, *Fingerprints of the Gods,* (Toronto, Ontario, Doubleday Canada Ltd., 1995)
36. Drunvalo Melchizadek, *Serpent of Light: The Movement of Earth's Kundalini and the Rise of the Female Light, 1949 to 2013,*(San Francisco, Red Wheel/Weiser, LLC, 2007)
37. Ibid
38. Carlos Barrios, *The Book of Destiny: Unlocking the Secrets of the Ancient Mayans and the Prophecy of 2012,* (New York, NY, HarperCollins Publishers, 2009)
39. Drunvalo Melchizadek, *Serpent of Light: The Movement of Earth's Kundalini and the Rise of the Female Light, 1949 to 2013,*(San Francisco, Red Wheel/Weiser, LLC, 2007
40. Carlos Barrios, *The Book of Destiny: Unlocking the Secrets of the Ancient Mayans and the Prophecy of 2012,* (New York, NY, HarperCollins Publishers, 2009)
41. www.sacredsites.com/americas/mexico/tula.html

42. Chris Morton and Ceri Louise Thomas, *The Mystery of the Crystal Skulls: Unlocking the Secrets of the Past, Present and Future*, (Rochester, VT, Bear & Company, 1997, 1998, 2002)
43. Adrian Gilbert and Maurice Cotterell, *The Mayan Prophecies: Unlocking the Secrets of a Lost Civilization*, (Shaftesbury, Dorset, UK, Element Books Ltd.,1995) citing John Mitchell, *Eccentric Lives and Peculiar Notions*, (London , Thames & Hudson, 1984) which cited Igantius Donnelly, *Atlantis the Ante-diluvian World*, revised by Egerton, Sykes, Sidgwick & Jackson, 1970
44. Patricia Mercier, *The Maya End Times: A Spiritual Adventure to the Heart of the Maya Prophecies for 2012*, (London, Watkins Publishing, 2008)
45. Chris Morton and Ceri Louise Thomas, *The Mystery of the Crystal Skulls: Unlocking the Secrets of the Past, Present and Future*, (Rochester, VT, Bear & Company, 1997, 1998, 2002.
46. Patricia Mercier, *The Maya End Times: A Spiritual Adventure to the Heart of the Maya Prophecies for 2012*, (London, Watkins Publishing, 2008)
47. Immanuel Velikovsky, *Earth in Upheaval* (New York, Pocket Books, Simon & Schuster, 1977)
48. Carlos Barrios, *The Book of Destiny: Unlocking the Secrets of the Ancient Mayans and the Prophecy of 2012*, (New York, NY, HarperCollins Publishers, 2009)
49. http://en.wikipedia.org/wiki/George_Clapp_Vaillant
50. Michael D. Coe, *Mexico: From the Olmecs to the Aztecs*, (London, Thames & Hudson, 1962). 2nd edition published by Thames & Hudson, NY, 1982, 4th edition 1994
51. Patricia Mercier, *The Maya End Times: A Spiritual Adventure to the Heart of the Maya Prophecies for 2012*, (London, Watkins Publishing, 2008)
52. Patricia Mercier, *The Maya End Times: A Spiritual Adventure to the Heart of the Maya Prophecies for 2012*, (London, Watkins Publishing, 2008)
53. Michael D. Coe, *Mexico: From the Olmecs to the Aztecs*, (London, Thames & Hudson, 1962). 2nd edition published by Thames & Hudson, NY, 1982, 4th edition 1994
54. David Freidel, Linda Schele, & Joy Parker, *Maya Cosmos: Three Thousand Years on the Shaman's Path*, (New York, NY, William Morrow and Company, Inc., 1995)

55. Barbara Hand Clow, *Catastrophobia: The Truth Behind Earth Changes in the Coming Age of Light,* (Rochester, VT, Bear and Company, 2001) pg. 142.
56. Graham Hancock, *Fingerprints of the Gods,* (Toronto, Ontario, Doubleday Canada Ltd., 1995)
57. Michael D. Coe, *Mexico: From the Olmecs to the Aztecs,* (London, Thames & Hudson, 1962, 2nd edition published by Thames & Hudson, NY, 1982, 4th edition 1994)
58. Translated by Dennis Tedlock, *Popol Vuh: The Mayan Book of the Dawn of Life,* (New York, Simon & Schuster, Inc., 1985)
59. http://en.wikipedia.org/wiki/Izapa
60. Barbara Tedlock, PH.D., *The Woman in the Shaman's Body: Reclaiming the Feminine in Religion and Medicine,* (New York, Bantam Dell, A Division of Random House, Inc., 2005)
61. Sandra Ingerman, *Soul Retrieval: Mending the Fragmented Self,* (New York, HarperCollins, 1991) citing Mircea Eliade, *Shamanism.*
62. Nan Moss with David Corbin, *Weather Shamanism: Harmonizing our Connection with the Elements,* (Rochester, VT, Bear & Company, 2008)
63. Geoff Stray, *Beyond 2012: Catastrophe or Awakening?* (Rochester, VT, Bear and Company, 2005, 2006, 2009)
64. David Freidel, Linda Schele, & Joy Parker, *Maya Cosmos: Three Thousand Years on the Shaman's Path,* (New York, NY, William Morrow and Company, Inc., 1995)
65. Ibid
66. John Major Jenkins, *Maya Cosmogenesis, 2012,* (Sante Fe, Bear & Company, 1998)
67. http://ambergriscaye.com/pages/mayan/mayasites.html
68. David Freidel, Linda Schele, & Joy Parker, *Maya Cosmos: Three Thousand Years on the Shaman's Path,* (New York, NY, William Morrow and Company, Inc., 1995)
69. Ibid
70. Carlos Barrios, *The Book of Destiny: Unlocking the Secrets of the Ancient Mayans and the Prophecy of 2012,* (New York, NY, HarperCollins Publishers, 2009)
71. David Freidel, Linda Schele, & Joy Parker, *Maya Cosmos: Three Thousand Years on the Shaman's Path,* (New York, NY, William Morrow and Company, Inc., 1995)
72. F. David Peat, *Blackfoot Physics: A Journey into the Native American Universe,* (Boston, Ma/York Beach, ME, Red Wheel/Weiser, 2002)

73. John Major Jenkins, *Maya Cosmogenesis, 2012*, (Sante Fe, Bear & Company, 1998) citing David Carrasco, *Religions of Mesoamerica:Cosmovision and Ceremonial Centers*, (New York, Harper & Row, 1990)
74. Ibid
75. Ibid
76. David Freidel, Linda Schele, & Joy Parker, *Maya Cosmos: Three Thousand Years on the Shaman's Path*, (New York, NY, William Morrow and Company, Inc., 1995)
77. Linda Schele and David Freidel, *A Forest of Kings: The Untold Story of the Ancient Maya*, (New York, William Morrow and Company, Inc., 1990)
78. http://wikipedia.org/wiki/Calakmul
79. Barbara Tedlock, Ph.D., *The Woman in the Shaman's Body: Reclaiming the Feminine in Religion and Medicine*, (New York, Bantam Dell, A Division of Random House, Inc., 2005)
80. Carl Johan Calleman, PH.D., "The Nine Underworlds: Expanding Levels of Consciousness," from *The Mystery of 2012: Predictions, Prophecies & Possibilities*, (Boulder, CO, Sounds True, Inc., 2007)
81. Patricia Mercier, *The Maya End Times: A Spiritual Adventure to the Heart of the Maya Prophecies for 2012*, (London, Watkins Publishing, 2008)
82. Chris Morton and Ceri Louise Thomas, *The Mystery of the Crystal Skulls: Unlocking the Secrets of the Past, Present and Future*, (Rochester, VT, Bear & Company, 1997, 1998, 2002)
83. Ibid
84. Ibid
85. Graham Hancock, *Fingerprints of the Gods*, (Toronto, Ontario, Doubleday Canada Ltd., 1995)
86. Anthony Aveni, *Stairways to the Stars: Skywatching in Three Ancient Cultures*, (New York, John Wiley & Sons, Inc., 1997)
87. Ibid
88. Brian Fagan, *From Black Land to Fifth Sun: The Science of Sacred Sites*, (Reading Massachusetts, Addison-Wesley, 1998)
89. Adrian Gilbert and Maurice Cotterell, *The Mayan Prophecies: Unlocking the Secrets of a Lost Civilization*, (Shaftesbury, Dorset, UK, Element Books Ltd.,1995)
90. Ibid
91. Graham Hancock, *Fingerprints of the Gods*, (Toronto, Ontario, Doubleday Canada Ltd., 1995)

92. Ibid citing Nigel Davis, *The Ancient Kingdoms of Mexico*, (London, Penguin Books, 1990)
93. Ibid
94. Adrian Gilbert and Maurice Cotterell, *The Mayan Prophecies: Unlocking the Secrets of a Lost Civilization*, (Shaftesbury, Dorset, UK, Element Books Ltd.,1995)
95. Michael D. Coe, *Mexico: From the Olmecs to the Aztecs*, (London, Thames & Hudson, 1962, 2nd edition published by Thames & Hudson, NY, 1982, 4th edition 1994)
96. Ibid
97. Ibid
98. Graham Hancock, *Fingerprints of the Gods*, (Toronto, Ontario, Doubleday Canada Ltd., 1995)
99. Michael D. Coe, *Mexico: From the Olmecs to the Aztecs*, (London, Thames & Hudson, 1962, 2nd edition published by Thames & Hudson, NY, 1982, 4th edition 1994)
100. Anthony Aveni, *Stairways to the Stars: Skywatching in Three Ancient Cultures,* (New York, John Wiley & Sons, Inc., 1997)
101. Drunvalo Melchizadek, *Serpent of Light: The Movement of Earth's Kundalini and the Rise of the Female Light, 1949 to 2013,*(San Francisco, Red Wheel/Weiser, LLC, 2007)
102. David Freidel, Linda Schele, & Joy Parker, *Maya Cosmos: Three Thousand Years on the Shaman's Path, (*New York, NY, William Morrow and Company, Inc., 1995)
103. John Major Jenkins, *Maya Cosmogenesis, 2012,* (Sante Fe, Bear & Company, 1998) citing David Carrasco, *Religions of Mesoamerica:Cosmovision and Ceremonial Centers, (*New York, Harper & Row, 1990)
104. Daniel Pinchbeck, *2012: The Return of Quetzalcoatl,* (New York, Jeremy P. Tarcher/Penguin, 2006)
105. Ibid

TIME AND THE MAYAN CALENDAR

1. David Freidel, Linda Schele, & Joy Parker, *Maya Cosmos: Three Thousand Years on the Shaman's Path, (*New York, NY, William Morrow and Company, Inc., 1995)
2. Graham Hancock, *Fingerprints of the Gods,* (Toronto, Ontario, Doubleday Canada Ltd., 1995)
3. Translated by Dennis Tedlock, *Popol Vuh: The Mayan Book of the Dawn of Life,* (New York, Simon & Schuster, Inc., 1985)

4. Carlos Barrios, *The Book of Destiny: Unlocking the Secrets of the Ancient Mayans and the Prophecy of 2012*, (New York, NY, HarperCollins Publishers, 2009)
5. Lawrence E. Joseph, *Apocalypse 2012: A Scientific Investigation into Civilization's End*, (New York, Morgan Road Books, an imprint of The Doubleday Broadway Publishing Group, a division of Random House, Inc., 2007)
6. Geoff Stray, *Beyond 2012: Catastrophe or Awakening?* (Rochester, VT, Bear and Company, 2005, 2006, 2009)
7. Ibid
8. Ibid
9. Translated by Dennis Tedlock, *Popol Vuh: The Mayan Book of the Dawn of Life*, (New York, Simon & Schuster, Inc., 1985)
10. Carlos Barrios, *The Book of Destiny: Unlocking the Secrets of the Ancient Mayans and the Prophecy of 2012*, (New York, NY, HarperCollins Publishers, 2009
11. John Major Jenkins, *The 2012 Story: The Myths, Fallacies, and Truth Behind the Most Intriguing Date in History*, (New York, Jeremy P. Tarcher/Penguin, 2009)
12. Ibid
13. Carlos Barrios, *The Book of Destiny: Unlocking the Secrets of the Ancient Mayans and the Prophecy of 2012*, (New York, NY, HarperCollins Publishers, 2009)
14. Ibid
15. Linda Schele and David Freidel, *A Forest of Kings: The Untold Story of the Ancient Maya*, (New York, William Morrow and Company, Inc., 1990)
16. Carlos Barrios, *The Book of Destiny: Unlocking the Secrets of the Ancient Mayans and the Prophecy of 2012*, (New York, NY, HarperCollins Publishers, 2009)
17. John Major Jenkins, *Maya Cosmogenesis, 2012*, (Sante Fe, Bear & Company, 1998) citing David Carrasco, *Religions of Mesoamerica:Cosmovision and Ceremonial Centers, (New York, Harper & Row, 1990)*
18. John Major Jenkins, *The 2012 Story: The Myths, Fallacies, and Truth Behind the Most Intriguing Date in History*, (New York, Jeremy P. Tarcher/Penguin, 2009)
19. Ibid
20. Ibid
21. Geoff Stray, *Beyond 2012: Catastrophe or Awakening?* (Rochester, VT, Bear and Company, 2005, 2006, 2009)

22. Barbara Hand Clow with Gerry Clow, *Alchemy of Nine Dimensions: The 2011/2012 Prophecies and Nine Dimensions of Consciousness*, (Charlottesville, VA, Hampton Roads Publishing Company, 2004, 2010)
23. Barbara Marciniak with Karen Marciniak and Tera Thomas, *Earth: Pleiadian Keys to the Living Library*, (Sante Fe, NM, Bear & Company, Publishing, 1995)
24. Ibid
25. Chris Morton and Ceri Louise Thomas, *The Mystery of the Crystal Skulls: Unlocking the Secrets of the Past, Present and Future*, (Rochester, VT, Bear & Company, 1997, 1998, 2002)
26. Anthony Aveni, *Stairways to the Stars: Skywatching in Three Ancient Cultures*, (New York, John Wiley & Sons, Inc., 1997)
27. Ibid
28. Geoff Stray, *Beyond 2012: Catastrophe or Awakening?* (Rochester, VT, Bear and Company, 2005, 2006, 2009)
29. Gregg Braden, *Fractal Time: The Secret of 2012 and a New World Age*, (Carlsbad, CA, Hay House, Inc., 2009) citing Terence McKenna, *The Invisible Landscape: Mind, Hallucinogens, and the I Ching*, (New York, HarperOne, 1994)
30. Ibid

ASTRONOMY AND ASTROLOGY

1. Gregory Sams, *Sun of gOd*, (San Francisco, Red Wheel/Weiser, LLC, 2009)
2. http://en.wikipedia.org/wiki/Plasma_(physics)
3. Gregory Sams, *Sun of gOd*, (San Francisco, Red Wheel/Weiser, LLC, 2009)
4. Patricia Daniels, *The New Solar System: Ice Worlds, Moons, and Planets Redefined*, (Washington, DC, National Geographic Society, 2009)
5. Gregory Sams, *Sun of gOd*, (San Francisco, Red Wheel/Weiser, LLC, 2009)
6. Adrian Gilbert and Maurice Cotterell, *The Mayan Prophecies: Unlocking the Secrets of a Lost Civilization*, (Shaftesbury, Dorset, UK, Element Books Ltd.,1995)
7. Patricia Daniels, *The New Solar System: Ice Worlds, Moons, and Planets Redefined*, (Washington, DC, National Geographic Society, 2009)
8. Ibid
9. Ibid
10. Patricia Daniels, *The New Solar System: Ice Worlds, Moons, and Planets Redefined*, (Washington, DC, National Geographic Society, 2009)

11. Geoff Stray, *Beyond 2012: Catastrophe or Awakening?* (Rochester, VT, Bear and Company, 2005, 2006, 2009)
12. Lawrence E. Joseph, *Apocalypse 2012: A Scientific Investigation into Civilization's End*, (New York, Morgan Road Books, an imprint of The Doubleday Broadway Publishing Group, a division of Random House, Inc., 2007)
13. Ibid
14. http://en.wikipedia.org/wiki/Solar_wind
15. Geoff Stray, *Beyond 2012: Catastrophe or Awakening?* (Rochester, VT, Bear and Company, 2005, 2006, 2009)
16. George A. Seielstad, *At the Heart of the Web: The Inevitable Genesis of Intelligent Life*, (Boston, Harcourt Brace Jovanovich, Inc., 1989)
17. Geoff Stray, *Beyond 2012: Catastrophe or Awakening?* (Rochester, VT, Bear and Company, 2005, 2006, 2009)
18. Ibid citing www.tmgnow.com/repository/global/planetophysical1.html
19. Carlos Barrios, *The Book of Destiny: Unlocking the Secrets of the Ancient Mayans and the Prophecy of 2012*, (New York, NY, HarperCollins Publishers, 2009)
20. *Science Week*, 2005.307:686
21. Ken Croswell, "Star Struck: Heart of the Milky Way", *National Geographic Magazine*, (Washington, DC, The National Geographic Society, December, 2010, Vol. 218, No. 6)
22. Geoff Stray, *Beyond 2012: Catastrophe or Awakening?* (Rochester, VT, Bear and Company, 2005, 2006, 2009) citing Paul LaViolette, *Earth Under Fire: Humanity's Survival of the Apocalypse*, (New York, Starlane Publications, 1997)
23. Barbara Hand Clow, *The Mayan Code: Time Acceleration and Awakening the World Mind*, (Rochester VT, Bear & Company, 2007) citing James Glanz, "Theorists Ponder a Cosmic Boost from Far, Far Away," *New York Times*, February 15, 2000.
24. http://science.nasa.gov/science-at-nasa/2010/27jul_spacequakes/
25. Geoff Stray, *2012 in Your Pocket*, (Virginia Beach, VA, A.R.E. Press, 2008)
26. Bruce H. Lipton, PH.D. and Steve Bhaerman, *Spontaneous Evolution: Our Positive Future (and a way to get there from here)*, (Carlsbad, CA, New York, Hay House, Inc., 2009)
27. Carlos Barrios, *The Book of Destiny: Unlocking the Secrets of the Ancient Mayans and the Prophecy of 2012*, (New York, NY, HarperCollins Publishers, 2009)
28. Ibid

29. Geoff Stray, "The Advent of the Post-Human Geo-Neuron," in Sounds True, Inc., *The Mystery of 2012: Predictions, Prophecies & Possibilities, edited by Tami Simon,* (Boulder, CO, Sounds True, Inc., 2007)
30. Gregory Sams, *Sun of gOd,* (San Francisco, Red Wheel/Weiser, LLC, 2009)
31. Annette Deyhle, Ph.D., Global Coherence Initiative, Boulder Creek, CA, communication to subscribers, October, 2009.
32. Adrian Gilbert and Maurice Cotterell, *The Mayan Prophecies: Unlocking the Secrets of a Lost Civilization,* (Shaftesbury, Dorset, UK, Element Books Ltd.,1995)
33. Lyn Birkbeck, *Understanding the Future: A Survivor's Guide to Riding the Cosmic Wave,* (London, Watkins Publishing, 2008)
34. Ibid
35. Ibid
36. Ibid
37. Edited by Daniel Pinchbeck and Ken Jordan, *Toward 2012: Perspectives on the Next Age,* (New York, The Penguin Group, 2008)
38. Ibid
39. Lawrence E. Joseph, *Apocalypse 2012: A Scientific Investigation into Civilization's End,* (New York, Morgan Road Books, an imprint of The Doubleday Broadway Publishing Group, a division of Random House, Inc., 2007)
40. Ibid
41. Ibid
42. Ibid
43. Ibid
44. Ibid
45. Barbara Marciniak with Karen Marciniak and Tera Thomas, *Earth: Pleiadian Keys to the Living Library,* (Sante Fe, NM, Bear & Company, Publishing, 1995)
46. Ibid
47. www.pbs.org/wgbh/nova/elegant/dimensions.html
48. www.fortunecity.com/emachines/e11/86/dimens.html
49. Ibid
50. Patricia Mercier, *The Maya End Times: A Spiritual Adventure to the Heart of the Maya Prophecies for 2012,* (London, Watkins Publishing, 2008)
51. Whitley Strieber, *The Key: A True Encounter,* (Copyright 2001, 2011 by Walker & Collier, Inc., Published by Jeremy P. Tarcher/Penguin, New York, 2011)

SIGNS AND PROPHECIES

1. James Redfield, *The Celestine Vision: Living the New Spiritual Awareness*, (New York, Warner Books, Inc., 1997)
2. Geoff Stray, *Beyond 2012: Catastrophe or Awakening?* (Rochester, VT, Bear and Company, 2005, 2006, 2009) citing Ken Carey, *Starseed*, (San Francisco, Harper, 1991), *The Starseed Transmissions* and *Visions*, (Uni-Sun, 1982, 1985)
3. Sounds True, Inc., *The Mystery of 2012: Predictions, Prophecies & Possibilities*, edited by Tami Simon, (Boulder, CO, Sounds True, Inc., 2007) citing Geoff Stray, "The Advent of the Post-Human Geo-Neuron"
4. http://www.near-death.com/forum/nde/000/05.html
5. Solara, *11:11 Inside the Doorway*, (Whitefish, MT, Star-Borne Unlimited, 1992)
6. http://www.near-death.com/experiences/reincarnation04.html
7. Carlos Barrios, *The Book of Destiny: Unlocking the Secrets of the Ancient Mayans and the Prophecy of 2012*, (New York, NY, HarperCollins Publishers, 2009)
8. Ibid
9. Sheila and Marcus Gillette, *The Soul Truth: A Guide to Inner Peace, The Teachings of THEO*, (Jeremy P. Tarcher/Penguin, Published by the Penguin Group, 2008)
10. Barbara Hand Clow with Gerry Clow, *Alchemy of Nine Dimensions: The 2011/2012 Prophecies and Nine Dimensions of Consciousness*, (Hampton Roads Publishing Company, Inc., 2004, 2010)
11. Ibid
12. David Ian Cowan, *Navigating the Collapse of Time*, (San Francisco, Red Wheel/Weiser, 2011)
13. Ibid
14. Geoff Stray, *Beyond 2012: Catastrophe or Awakening?* (Rochester, VT, Bear and Company, 2005, 2006, 2009)
15. Gregg Braden, *Awakening to Zero Point: The Collective Initiation*, (Bellevue, WA, Radio Bookstore Press, 1997)
16. F. David Peat, *Blackfoot Physics: A Journey into the Native American Universe*, (Boston, Ma/York Beach, ME, Red Wheel/Weiser, 2002)
17. Ibid
18. Nan Moss with David Corbin, *Weather Shamanism: Harmonizing Our Connection with the Elements*, (Rochester, VT, Bear & Company, 2008)

19. Bruce H. Lipton, PH.D. and Steve Bhaerman, *Spontaneous Evolution: Our Positive Future (and a way to get there from here),* (Carlsbad, CA, New York, Hay House, Inc., 2009)
20. Ibid
21. Hazel Courteney, *Countdown to Coherence: A Spiritual Journey Toward a Scientific Theory of Everything,* (London, UK, Watkins Publishing, 2010)
22. F. David Peat, *Blackfoot Physics: A Journey into the Native American Universe,* (Boston, MA/York Beach, ME, Red Wheel/Weiser, 2002)
23. Míceál Ledwith, D.D., LL.D. & Klaus Heinemann, Ph.D., *The Orb Project,* (New York/London, Atria Books, A Division of Simon & Schuster, 2007) with Foreword by Dr. William A. Tiller
24. Hazel Courteney, *Countdown to Coherence: A Spiritual Journey Toward a Scientific Theory of Everything,* (London, Watkins Publishing, 2010, Distributed in the USA and Canada by Sterling Publishing Co., Inc. New York)
25. Ibid
26. Ibid
27. Klaus Heinemann, Ph.D, and Gundi Heinemann, *Orbs: Their Mission and Messages of Hope,* (Carlsbad, CA/New York, Hay House, Inc., 2010)
28. Ibid
29. Geoff Stray, *Beyond 2012: Catastrophe or Awakening?* (Rochester, VT, Bear and Company, 2005, 2006, 2009)
30. Freddy Silva, *Secrets in the Fields: The Science and Mysticism of Crop Circles,* (Charlottesville, VA, Hampton Roads Publishing Company, Inc., 2002)
31. Ibid
32. Ibid
33. Ibid
34. Ibid
35. Ibid
36. http://www.temporarytemples.co.uk/faqs/
37. Geoff Stray, *Beyond 2012: Catastrophe or Awakening?* (Rochester, VT, Bear and Company, 2005, 2006, 2009)
38. Ibid
39. Ibid
40. John Major Jenkins, *The 2012 Story: The Myths, Fallacies, and Truth Behind the Most Intriguing Date in History,* (New York, Jeremy P. Tarcher/Penguin, 2009)
41. Ibid

42. Ibid
43. Barbara Hand Clow, *Catastrophobia:The Truth Behind Earth Changes in the Coming Age of Light*, (Rochester, VT, Bear & Company, 2001)
44. Gregg Braden, *The Isaiah Effect: Decoding the Lost Science of Prayer and Prophecy*, (New York, Harmony Books/Random House, Inc., 2000)
45. Ibid
46. Sounds True, Inc., *The Mystery of 2012: Predictions, Prophecies & Possibilities, edited by Tami Simon*, (Boulder, CO, Sounds True, Inc., 2007) citing Geoff Stray, "The Advent of the Post-Human Geo-Neuron"
47. Geoff Stray, *Beyond 2012: Catastrophe or Awakening?* (Rochester, VT, Bear and Company, 2005, 2006, 2009) citing Frank Waters, *Book of the Hopi*, (New York, Ballantine Books, 1969)
48. Sandra Ingerman, *Soul Retrieval: Mending the Fragmented Self*, (New York, HarperCollins, 1991) citing Maxwell C. Cade and Nona Coxhead, *The Awakened Mind: Biofeedback and the Development of Higher States of Awareness*, (Great Britain, Element Books, 1979) and Izhak Bentov, *Stalking the Wild Pendulum*, (New York, E.F. Dutton, 1977)
49. Geoff Stray, *Beyond 2012: Catastrophe or Awakening?* (Rochester, VT, Bear and Company, 2005, 2006, 2009)
50. Gregg Braden, *The Divine Matrix: Bridging Time, Space, Miracles and Belief*, (Carlsbad, CA, Hay House, Inc., 2007)
51. Ibid
52. Brian Fagan, *From Black Land to Fifth Sun: The Science of Sacred Sites*, (Reading, MA, Helix Books/ Addison-Wesley, 1998) citing Michael Zeilik, *Archeoastronomy 8:S1-24*, "The Ethnoastronomy of the Historic Pueblos, I: Calendrical Sun Watching," 1985
53. Karen Alexander, http://www.temporarytemples.co.uk/blog/crop-circles-sound-bubbles/
54. Patricia Mercier, *The Maya End Times: A Spiritual Adventure to the Heart of the Maya Prophecies for 2012*, (London, Watkins Publishing, 2008)
55. Ibid
56. Graham Hancock, *Fingerprints of the Gods*, (Toronto, Ontario, Doubleday Canada Ltd., 1995)
57. Ron Redfern, *Corridors of Time: 1,700,000,000 Years of Earth at Grand Canyon*, (New York, Times Books, 1980)
58. Nan Moss with David Corbin, *Weather Shamanism: Harmonizing Our Connection with the Elements*, (Rochester, VT, Bear & Company, 2008)

59. Ibid
60. Edited by Daniel Pinchbeck and Ken Jordan, *Toward 2012: Perspectives on the Next Age*, (New York, Jeremy P. Tarcher/Penguin, 2008) in chapter by John Major Jenkins, "Mayan Shamanism and 2012: A Psychedelic Cosmology" quoting Gerardo Reichel-Dolmatoff, "Center of Day," 1982.
61. F. David Peat, *Blackfoot Physics: A Journey into the Native American Universe*, (Boston, Ma/York Beach, ME, Red Wheel/Weiser, 2002)
62. Ibid
63. Freddy Silva, *Secrets in the Fields: The Science and Mysticism of Crop Circles*, (Charlottesville, VA, Hampton Roads Publishing Company, Inc., 2002) citing Credo Mutwa, *Isilwane: The Animal*, (Cape Town, S. Africa, Struick, 1996)
64. Freddy Silva, *Secrets in the Fields: The Science and Mysticism of Crop Circles*, (Charlottesville, VA, Hampton Roads Publishing Company, Inc., 2002)
65. Sounds True, Inc., *The Mystery of 2012: Predictions, Prophecies & Possibilities, edited by Tami Simon*, (Boulder, CO, Sounds True, Inc., 2007) citing from chapter by Christine Page, M.D., "A Time to Remember"
66. Ibid
67. Patricia Mercier, *The Maya End Times: A Spiritual Adventure to the Heart of the Maya Prophecies for 2012*, (London, Watkins Publishing, 2008)
68. Anthony Aveni, *Stairways to the Stars: Skywatching in Three Ancient Cultures*, (New York, John Wiley & Sons, Inc., 1997)
69. Patricia Mercier, *The Maya End Times: A Spiritual Adventure to the Heart of the Maya Prophecies for 2012*, (London, Watkins Publishing, 2008)
70. Ibid
71. Ibid
72. Geoff Stray, *Beyond 2012: Catastrophe or Awakening?* (Rochester, VT, Bear and Company, 2005, 2006, 2009) from Elizabeth B. Jenkins, *A Woman's Spiritual Adventure in the Heart of the Andes*, (London, Piatkus, 1997)
73. Geoff Stray, *Beyond 2012: Catastrophe or Awakening?* (Rochester, VT, Bear and Company, 2005, 2006, 2009)
74. Alberto Villoldo and Erik Jendresen, *Dance of the Four Winds: Secrets of the Inca Medicine Wheel*, (Rochester, VT, Destiny Books, 1990, 1995)

75. Alberto Villoldo and Erik Jendresen, *Island of the Sun: Mastering the Inca Medicine Wheel*, (Rochester, VT, Destiny Books, 1992, 1995)
76. Ibid
77. Ibid
78. Alberto Villoldo, Ph.D. and Stanley Krippner, Ph.D., *Healing States: A Journey into the World of Spiritual Healing and Shamanism*, (New York, A Fireside Book, Simon & Schuster, Inc., 1987)
79. Alberto Villoldo and Erik Jendresen, *Dance of the Four Winds: Secrets of the Inca Medicine Wheel*, (Rochester, VT, Destiny Books, 1990, 1995)
80. Alberto Villoldo, Ph.D. and Stanley Krippner, Ph.D., *Healing States: A Journey into the World of Spiritual Healing and Shamanism*, (New York, A Fireside Book, Simon & Schuster, Inc., 1987)
81. Freddy Silva, *Secrets in the Fields: The Science and Mysticism of Crop Circles*, (Charlottesville, VA, Hampton Roads Publishing Company, Inc., 2002) citing John Anthony West, *Serpent in the Sky*, (Wheaton IL, Quest, 1993)
82. Ibid
83. Geoff Stray, *Beyond 2012: Catastrophe or Awakening?* (Rochester, VT, Bear and Company, 2005, 2006, 2009) quoting Jay Weidner of the film, *Healing the Luminous Body: The Way of the Shaman*, www.sacredmysteries.com
84. Graham Hancock, *Fingerprints of the Gods*, (Toronto, Ontario, Doubleday Canada Ltd., 1995)
85. Ibid
86. Ibid
87. http://www.oprah.com/spirit/Spirtual-Healer-John-of-God-Susan-Casey
88. Míceál Ledwith, D.D., LL.D. & Klaus Heinemann, Ph.D., *The Orb Project*, (New York/London, Atria Books, A Division of Simon & Schuster, 2007) with Foreword by Dr. William A. Tiller
89. Ibid
90. Lawrence E. Joseph, *Apocalypse 2012: A Scientific Investigation into Civilization's End*, (New York, Morgan Road Books, an imprint of The Doubleday Broadway Publishing Group, a division of Random House, Inc., 2007)
91. Solara Antara Amaa-Ra, *The Star-Borne: A Remembrance for the Awakened Ones*, (Charlottesville, VA, Star-Borne Unlimited, 1989, 1991)

92. Patricia Mercier, *The Maya End Times: A Spiritual Adventure to the Heart of the Maya Prophecies for 2012*, (London, Watkins Publishing, 2008)
93. Ibid
94. Ibid
95. Ibid
96. Ibid
97. Ibid
98. Chris Morton and Ceri Louise Thomas, *The Mystery of the Crystal Skulls: Unlocking the Secrets of the Past, Present, and Future*, (Rochester, VT, Bear & Company, 1997, 1998, 2002)
99. Ibid
100. Ibid
101. Ibid
102. Ibid
103. Ibid
104. Ibid
105. Gregg Braden, *The Isaiah Effect: Decoding the Lost Science of Prayer and Prophecy*, (New York, Harmony Books/Random House, Inc., 2000)
106. Ibid citing Michael Drosnin, *The Bible Code*, (New York, Simon & Schuster, 1997)
107. Ibid citing Jack Cohen and Ian Stewart, *The Collapse of Chaos*, (New York, Penguin Books, 1994)
108. Ibid citing Michael Drosnin, *The Bible Code*, (New York, Simon & Schuster, 1997)
109. Ibid
110. Sylvia Browne with Lindsay Harrison, *Phenomenon: Everything You Need to Know About the Paranormal*, (New York, Dutton, a member of the Penguin Group (USA) inc., 2005)
111. Ibid
112. Sounds True, Inc., *The Mystery of 2012: Predictions, Prophecies & Possibilities*, edited by Tami Simon, (Boulder, CO, Sounds True, Inc., 2007) citing Carl Johan Calleman, Ph.D., chapter "The Nine Underworlds"
113. James Redfield, Michael Murphy and Sylvia Timbers, *God and the Evolving Universe: The Next Step in Personal Evolution*, (New York, Jeremy P. Tarcher/Putnam, 2002)
114. Gregg Braden, *The Divine Matrix: Bridging Time, Space, Miracles and Belief*, (Carlsbad, CA, Hay House, Inc., 2007)
115. www.Abraham-Hicks.com

116. Geoff Stray, *Beyond 2012: Catastrophe or Awakening?* (Rochester, VT, Bear and Company, 2005, 2006, 2009)
117. http://www.diagnosis2012.co.uk
118. Lazaris, *Lazaris: The Sirius Connection*, (Orlando, FL, NPN Publishing, Inc., 1996)
119. Ibid
120. Geoff Stray, *Beyond 2012: Catastrophe or Awakening?* (Rochester, VT, Bear and Company, 2005, 2006, 2009)
121. Ibid
122. Sylvia Brown with Lindsay Harrison, *End of Days: Predictions and Prophecies about the End of the World*, (New York, Penguin Group, 2008)
123. Ibid
124. TUT Enterprises, Inc., *2012 A Wrinkle in Time*, featuring and produced by Mike Dooley, (TUT Enterprises, Inc., 2010)

CONSCIOUSNESS

1. John Major Jenkins, *The 2012 Story: The Myths, Fallacies, and Truth Behind the Most Intriguing Date in History*, (New York, Jeremy P. Tarcher/Penguin, 2009)
2. Ibid
3. http://en.wikipedia.org/wiki/New_Thought
4. http://www.livinglifefully.com/people/ralphwaldotrine.htm
5. James Allen, *As You Think: Become the Master of Your Own Destiny*, edited version by Marc Allen, (San Rafael, CA, New World Library, 1987)
6. Lynne McTaggart, *The Intention Experiment: Using Your Thoughts to Change Your Life and the World*, (New York, Free Press, A Division of Simon & Schuster, Inc., 2007)
7. Dr. Wayne W. Dyer, *You'll See It When You Believe It*, (New York, Avon Books, 1989)
8. Ruby Nelson, *The Door of Everything*, (Marina Del Rey, CA, DeVorss & Co., Publishers, 1963)
9. Gregg Braden, *Awakening to Zero Point: The Collective Initiation*, (Bellevue, WA, Radio Bookstore Press, 1993, 1994, 1997)
10. Ken Carey, *Vision*, (Kansas City, MO, UNI*/SUN, 1985) from the Introduction by Jean Houston, Ph.D.
11. Ibid
12. Lyn Birkbeck, *Understanding the Future: A Survivor's Guide to Riding the Cosmic Wave*, (London, Watkins Publishing, 2008)

13. Gregory Sams, *Sun of gOd: Consciousness and the Self-Organizing Force that Underlies Everything*, (San Francisco, Red Wheel/Weiser, LLC., 2009)
14. Míceál Ledwith, D.D., LL.D. & Klaus Heinemann, Ph.D., *The Orb Project*, (New York/London, Atria Books, A Division of Simon & Schuster, 2007)
15. Sounds True, Inc., *The Mystery of 2012: Predictions, Prophecies & Possibilities, edited by Tami Simon,* (Boulder, CO, Sounds True, Inc., 2007) citing from chapter by Meg Blackburn Losey, Ph.D., "2012 Awakening to Greater Reality"
16. Gregory Sams, *Sun of gOd: Consciousness and the Self-Organizing Force that Underlies Everything*, (San Francisco, Red Wheel/Weiser, LLC., 2009)
17. Eckhart Tolle, *Stillness Speaks*, (Novato, CA and Vancouver, Canada, New World Library/Namaste Publishing, 2003)
18. Eckhart Tolle, *A New Earth: Awakening to Your Life's Purpose*, (New York, Dutton, Penguin Group (USA) Inc., 2005 – a Namaste Publishing Book)
19. Fred Alan Wolfe, Ph.D., *Mind Into Matter: A New Alchemy of Science and Spirit,* (Needham, MA, Moment Point Press, Inc., 2001)
20. Danaan Parry, *The Essene Book of Days: 1997*, (Bainbridge Island, WA, Earthstewards Network, 1996)
21. Daniel Pinchbeck, *2012: The Return of Quetzalcoatl*, (New York, Jeremy P. Tarcher/Penguin, 2006)
22. Edited by Daniel Pinchbeck and Ken Jordan, *Toward 2012: Perspectives on the Next Age*, (New York, Jeremy P. Tarcher/Penguin, 2008) citing from chapter "Gnosis: The Not-So-Secret History of Jesus" by Jonathan Phillips
23. http://www/onenessuniversity.org/about_us_history.html
24. Ibid
25. Arjuna Ardagh, *Leap Before you Look: 72 Shortcuts for Getting Out of your Mind and into the Moment*, (Boulder, CO, Sounds True, Inc., 2008) from 'About the Author'
26. Bruce H. Lipton, PH.D. and Steve Bhaerman, *Spontaneous Evolution: Our Positive Future (and a way to get there from here)*, (Carlsbad, CA, New York, Hay House, Inc., 2009)
27. Lawrence E. Joseph, *Apocalypse 2012: A Scientific Investigation into Civilization's End*, (New York, Morgan Road Books, an imprint of The Doubleday Broadway Publishing Group, a division of Random House, Inc., 2007)

28. Mary J. Shomon, *The Menopause Thyroid Solution: Overcome Menopause by Solving Your Hidden Thyroid Problems*, (New York, HarperCollins Publishers, 2009)
29. www.onespirit.com
30. John J. Liptak, Ed.D., *2012: Catalyst for Your Spiritual Awakening: Using the Mayan Tree of Life to Discover Your Higher Purpose*, (Woodbury, MN, Llewellyn Publications, 2010)
31. Ibid
32. Kenneth X. Carey, *The Starseed Transmissions: An Extraterrestrial Report*, (Kansas City, MO, Uni*Sun, 1982)
33. Barbara Hand Clow, *The Mayan Code: Time Acceleration and Awakening the World Mind*, (Rochester, VT, 2007)
34. Edited by Martine Vallée, *The Great Shift: Co-creating a New World for 2012 and Beyond*, (San Francisco, CA/Newburyport, MA, Red Wheel/Weiser Books, 2009)
35. Hal Zina Bennett, *The Lens of Perception: A User's Guide to Higher Consciousness*, (Berkeley, CA, Celestial Arts, imprint of Ten Speed Press, 1987, 1994, 2007)
36. Layne Redmond, *When the Drummers were Women: A Spiritual History of Rhythm*, (New York, Three Rivers Press/Crown Publishing Group, 1997)
37. Vianna Stibal, *Theta Healing™:Introducing an Extraordinary Energy Healing Modality*, (Carlsbad, CA, Hay House, Inc., 2010)
38. Edited by Daniel Pinchbeck and Ken Jordan, *Toward 2012: Perspectives on the Next Age*, (New York, The Penguin Group, 2008) citing Stanislav Grof, "A New Understanding of the Psyche"
39. Barbara Marciniak with Karen Marciniak and Tera Thomas, *Earth: Pleiadian Keys to the Living Library*, (Sante Fe, NM, Bear & Company, Publishing, 1995)
40. Lynne McTaggart, *The Intention Experiment: Using Your Thoughts to Change Your Life and the World*, (New York, Free Press, A Division of Simon & Schuster, Inc., 2007)
41. Ibid
42. Ibid
43. Ibid
44. Ibid
45. Ibid
46. Ibid
47. Ibid

48. Esther and Jerry Hicks: The Teachings of Abraham, *Getting Into the Vortex: Guided Meditations CD and User Guide*, (Carlsbad, CA/New York, Hay House, Inc. 2010)
49. Lynne McTaggart, *The Intention Experiment: Using Your Thoughts to Change Your Life and the World*, (New York, Free Press, A Division of Simon & Schuster, Inc., 2007)
50. Ibid
51. Ibid
52. Ibid
53. Dr. Wayne W. Dyer, *The Power of Intention: Learning to Co-create Your World Your Way*, (Carlsbad, CA, Hay House, Inc., 2004)
54. Ibid
55. Gregg Braden, *The Divine Matrix: Bridging Time, Space, Miracles and Belief*, (Carlsbad, CA, Hay House, Inc., 2007)
56. Ibid
57. Gregory Sams, *Sun of gOd: Consciousness and the Self-Organizing Force that Underlies Everything*, (San Francisco, Red Wheel/Weiser, LLC., 2009)
58. Gregg Braden, *The Divine Matrix: Bridging Time, Space, Miracles and Belief*, (Carlsbad, CA, Hay House, Inc., 2007)
59. Ibid
60. Ibid
61. Ibid
62. Ibid citing Coleman Barks, *The Illuminated Rumi*, (New York, Broadway Books, 1997)
63. Gregg Braden, *The Spontaneous Healing of Belief: Shattering the Paradigm of False Limits*, (Carlsbad, CA, Hay House, Inc., 2008)
64. Gregg Braden, *The Divine Matrix: Bridging Time, Space, Miracles and Belief*, (Carlsbad, CA, Hay House, Inc., 2007)
65. Ibid
66. Sounds True, Inc., *The Mystery of 2012: Predictions, Prophecies & Possibilities*, edited by Tami Simon, (Boulder, CO, Sounds True, Inc., 2007) citing Arjuna Ardagh, in chapter "The Clock is Ticking"
67. Ibid
68. Esther and Jerry Hicks, (The Teachings of Abraham), *The Law of Attraction: The Basics of the Teachings of Abraham™*, (Carlsbad, CA, Hay House, Inc., 2006)
69. Ibid
70. Ibid
71. Ibid

72. Gregg Braden, *The Divine Matrix: Bridging Time, Space, Miracles and Belief,* (Carlsbad, CA, Hay House, Inc., 2007)
73. Esther and Jerry Hicks, (The Teachings of Abraham), *The Law of Attraction: The Basics of the Teachings of Abraham™*, (Carlsbad, CA, Hay House, Inc., 2006)
74. Esther and Jerry Hicks (The Teachings of Abraham), *The Amazing Power of Deliberate Intent: Living the Art of Allowing,* ((Carlsbad, CA, Hay House, Inc., 2006)
75. Esther and Jerry Hicks (The Teachings of Abraham), *Ask and It Is Given: Learning to Manifest Your Desires,* (Carlsbad, CA, Hay House, Inc., 2004
76. Sounds True, Inc., *The Mystery of 2012: Predictions, Prophecies & Possibilities, edited by Tami Simon,* (Boulder, CO, Sounds True, Inc., 2007) citing Gill Edwards, in chapter "Wild Love Sets Us Free"
77. Ibid, citing Esther and Jerry Hicks (The Teachings of Abraham), *Ask and It Is Given: Learning to Manifest Your Desires,* (Carlsbad, CA, Hay House, Inc., 2004)
78. Gill Edwards, *Wild Love,* (London, Piatkus Books, 2006)
79. Sounds True, Inc., *The Mystery of 2012: Predictions, Prophecies & Possibilities, edited by Tami Simon,* (Boulder, CO, Sounds True, Inc., 2007) citing Gill Edwards, in chapter "Wild Love Sets Us Free"
80. Alberto Villoldo, Ph.D. and Stanley Krippner, Ph.D., *Healing States: A Journey Into the World of Spiritual Healing and Shamanism,* (New York, A Fireside Book, Simon & Schuster, Inc., 1986, 1987)
81. Ibid
82. Alberto Villoldo, Ph.D., *Illumination: The Shaman's Way of Healing,* (Carlsbad, CA, Hay House, Inc., 2010)
83. Ibid
84. Ibid
85. Edited by Stephen Mitchell, *The Essence of Wisdom: Words from the Masters to Illuminate the Spiritual Path,* (New York, Broadway Books, 1998)
86. Sounds True, Inc., *The Mystery of 2012: Predictions, Prophecies & Possibilities, edited by Tami Simon,* (Boulder, CO, Sounds True, Inc., 2007) citing John Major Jenkins, in chapter "The Origins of the 2012 Revelation"
87. Daniel Pinchbeck, *2012: The Return of Quetzalcoatl,* (New York, Jeremy P. Tarcher/Penguin, 2006)
88. www.Abraham-Hicks.com
89. Gregg Braden, *The Spontaneous Healing of Belief: Shattering the Paradigm of False Limits,* (Carlsbad, CA, Hay House, Inc., 2008)

90. Ibid
91. Sounds True, Inc., *The Mystery of 2012: Predictions, Prophecies & Possibilities*, edited by Tami Simon, (Boulder, CO, Sounds True, Inc., 2007) citing James O'Dea, in chapter "You Were Born for Such a Time as This"
92. Bruce H. Lipton, PH.D. and Steve Bhaerman, *Spontaneous Evolution: Our Positive Future (and a way to get there from here)*, (Carlsbad, CA, New York, Hay House, Inc., 2009) citing Aura Glaser, *A Call to Compassion: Bringing Buddhist Practices of the Heart into the Soul of Psychology*, (Berwick, ME, Nicholas-Hays, 2005)
93. Vimala Thakar, Vimala Programs California, *The Eloquence of Living: Meeting Life with Freshness, Fearlessness, and Compassion*, (San Rafael, CA, New World Library, 1989)
94. Whitley Strieber, *The Key: A True Encounter*, (Copyright 2001, 2011 by Walker & Collier, Inc., Published by Jeremy P. Tarcher/Penguin, New York, 2011)
95. www.mooji.org/biography.html
96. Whitley Strieber, *The Key: A True Encounter*, (Copyright 2001, 2011 by Walker & Collier, Inc., Published by Jeremy P. Tarcher/Penguin, New York, 2011)
97. Sounds True, Inc., *The Mystery of 2012: Predictions, Prophecies & Possibilities*, edited by Tami Simon, (Boulder, CO, Sounds True, Inc., 2007) citing John L. Peterson, in chapter "Getting to 2012"
98. Sounds True, Inc., *The Mystery of 2012: Predictions, Prophecies & Possibilities*, edited by Tami Simon, (Boulder, CO, Sounds True, Inc., 2007) citing Llewellyn Vaughan-Lee, Ph.D., in chapter "An Awakening World"
99. Sounds True, Inc., *The Mystery of 2012: Predictions, Prophecies & Possibilities*, edited by Tami Simon, (Boulder, CO, Sounds True, Inc., 2007) citing Christine Page, M.D., in chapter "A Time to Remember"
100. Nan Moss with David Corbin, *Weather Shamanism: Harmonizing our Connection with the Elements*, (Rochester, VT, Bear & Company, 2008)
101. Ibid
102. Neale Donald Walsch, *Mother of Invention: The Legacy of Barbara Marx Hubbard*, (Carlsbad, CA, Hay House, Inc., 2011)
103. Alberto Villoldo and Erik Jendresen, *Dance of the Four Winds: Secrets of the Inca Medicine Wheel*, (Rochester, VT, Destiny Books, 1990, 1995)

104. Edited by Daniel Pinchbeck and Ken Jordan, *Toward 2012: Perspectives on the Next Age*, (New York, Jeremy P. Tarcher/Penguin, 2008) in chapter by Alberto Villoldo, "Jaguar Medicine"
105. Hunbatz Men, *Secrets of Mayan Science/Religion*, translated by Diana Gubiseh Ayala & James Jennings Dunlap II, (Sante Fe, NM, Bear & Company Publishing, 1990)
106. Whitley Strieber, *The Key: A True Encounter*, (Copyright 2001, 2011 by Walker & Collier, Inc., Published by Jeremy P. Tarcher/Penguin, New York, 2011)
107. Dr. Wayne W. Dyer, *The Power of Intention: Learning to Co-create Your World Your Way*, (Carlsbad, CA, Hay House, Inc., 2004)
108. http://en.wikipedia.org/wiki/Chris_Griscom
109. Christina Griscom, *Ecstasy is a New Frequency: Teachings of the Light Institute*, (Santa Fe, NM, Bear & Company, 1987)
110. www.redelk.net/website/CrystalKullMsg.htm
111. Daniel Pinchbeck, *2012: The Return of Quetzalcoatl*, (New York, Jeremy P. Tarcher/Penguin, 2006)
112. Ibid
113. Eckhart Tolle, *Stillness Speaks*, (Novato, CA/Vancouver, Canada, New World Library/Namaste Publishing, 2003)
114. Carlos Barrios, *The Book of Destiny: Unlocking the Secrets of the Ancient Mayans and the Prophecy of 2012*, (New York, NY, HarperCollins Publishers, 2009)
115. Sheila and Marcus Gillette, *The Soul Truth: A Guide to Inner Peace, The Teachings of THEO*, (Jeremy P. Tarcher/Penguin, Published by the Penguin Group, 2008)
116. Geoff Stray, *Beyond 2012: Catastrophe or Awakening?* (Rochester, VT, Bear and Company, 2005, 2006, 2009) citing Adrian Gilbert, *Signs in the Sky: Prophecies for the Birth of a New Age*, (London, Bantam, 2000)
117. Giorgio De Santillana & Hertha Von Dechend, *Hamlet's Mill: An Essay Investigating the Origins of Human Knowledge and Its Transmission Through Myth*, (Jaffrey, NH, David R. Grodine, Publisher, Inc., 1977)
118. Chris Morton and Ceri Louise Thomas, *The Mystery of the Crystal Skulls: Unlocking the Secrets of the Past, Present, and Future*, (Rochester, VT, Bear & Company, 1997, 1998, 2002)
119. Sylvia Browne with Lindsay Harrison, *Phenomenon: Everything You Need to Know About the Paranormal*, (New York, Dutton, a member of the Penguin Group (USA) inc., 2005)
120. Ibid

121. Eckhart Tolle, *A New Earth: Awakening to Your Life's Purpose*, (New York, Dutton, Penguin Group (USA) Inc., 2005)
122. Sounds True, Inc., *The Mystery of 2012: Predictions, Prophecies & Possibilities, edited by Tami Simon,* (Boulder, CO, Sounds True, Inc., 2007) citing Carl Johan Calleman, Ph.D. in chapter "The Nine Underworlds"
123. www.abraham-hicks.com
124. Whitley Strieber, *The Key: A True Encounter,* (Copyright 2001, 2011 by Walker & Collier, Inc., Published by Jeremy P. Tarcher/Penguin, New York, 2011)
125. Sounds True, Inc., *The Mystery of 2012: Predictions, Prophecies & Possibilities, edited by Tami Simon,* (Boulder, CO, Sounds True, Inc., 2007) citing Daniel Pinchbeck, in chapter "How the Snake Sheds Its Skin"
126. Ibid
127. Sounds True, Inc., *The Mystery of 2012: Predictions, Prophecies & Possibilities, edited by Tami Simon,* (Boulder, CO, Sounds True, Inc., 2007) citing Jean Houston, Ph.D., in chapter "Jump Time is Now"
128. George A. Seielstad, *At the Heart of the Web: The Inevitable Genesis of Intelligent Life,* (Boston, Harcourt Brace Jovanovich, Inc., 1989)
129. Gregg Braden, *Awakening to Zero Point: The Collective Initiation,* (Bellevue, WA, Radio Bookstore Press, 1993, 1994, 1997)
130. Alberto Villoldo, Ph.D., *Courageous Dreaming: How Shamans Dream the World into Being,* (Carlsbad, CA, Hay House, Inc., 2008)
131. Patricia Mercier, *The Maya End Times: A Spiritual Adventure to the Heart of the Maya Prophecies for 2012,* (London, Watkins Publishing, 2008)
132. Ibid
133. http://en.wikipedia.org/wiki/Pierre_Teilhard_de_Chardin
134. Geoff Stray, *2012 In Your Pocket,* (First A.R.E. Press Edition, March 2009) © 2008 by Geoff Stray
135. John Major Jenkins, *The 2012 Story: The Myths, Fallacies, and Truth Behind the Most Intriguing Date in History,* (New York, Jeremy P. Tarcher/Penguin, 2009)
136. James Redfield, Michael Murphy and Sylvia Timbers, *God and the Evolving Universe: The Next Step in Personal Evolution,* (New York, Jeremy P. Tarcher/Putnam, 2002)
137. Ibid
138. Alberto Villoldo, Ph.D., *Yoga, Power, and Spirit: Patanjali: the Shaman,* (Carlsbad, CA, Hay House, Inc., 2007)

139. Wayne W. Dyer, *There's A Spiritual Solution to Every Problem*, (New York, Harper/Collins Publishers, Inc., 2001)
140. Edited by Daniel Pinchbeck and Ken Jordan, *Toward 2012: Perspectives on the Next Age*, (New York, Jeremy P. Tarcher/Penguin, 2008) in chapter "Impossible Dreams" by Stephen Duncombe
141. Daniel Pinchbeck, *2012: The Return of Quetzalcoatl*, (New York, Jeremy P. Tarcher/Penguin, 2006)
142. Bruce H. Lipton, PH.D. and Steve Bhaerman, *Spontaneous Evolution: Our Positive Future (and a way to get there from here)*, (Carlsbad, CA, New York, Hay House, Inc., 2009)
143. Neale Donald Walsch, *Mother of Invention: The Legacy of Barbara Marx Hubbard*, (Carlsbad, CA, Hay House, Inc., 2011)
144. Ibid
145. Ibid
146. Ibid
147. www.noosphere.princeton.edu
148. Ibid
149. Jonathan Goldman, *The Divine Name: The Sound That Can Change the World*, (Carlsbad, CA, Hay House, Inc., 2010)
150. John Major Jenkins, *Maya Cosmogenesis, 2012*, (Sante Fe, Bear & Company, 1998)
151. Whitley Strieber, *The Key: A True Encounter*, (Copyright 2001, 2011 by Walker & Collier, Inc., Published by Jeremy P. Tarcher/Penguin, New York, 2011)
152. Geoff Stray, *Beyond 2012 Catastrophe or Awakening: A Complete Guide to End-of-Time Predictions*, (Rochester, VT, Bear & Company, 2005, 2006, 2009)
153. Alberto Villoldo, Ph.D., *Mending the Past and Healing the Future with Soul Retrieval*, (Carlsbad, CA, Hay House, Inc., 2005)
154. Alberto Villoldo, *The Four Insights: Wisdom, Power, and Grace of the Earthkeepers*, (Carlsbad, CA, Hay House, Inc., 2006)
155. James Redfield, *The Tenth Insight: Holding the Vison*, (New York, Warner Books, Inc., 1996)
156. Drunvalo Melchizadek, *Serpent of Light: The Movement of Earth's Kundalini and the Rise of the Female Light, 1949 to 2013*, (San Francisco, Red Wheel/Weiser, LLC, 2007)
157. Barbara Hand Clow, *Catastrophobia: The Truth Behind Earth Changes in the Coming Age of Light*, (Rochester, VT, Bear & Company, 2001
158. Gregg Braden, *The Isaiah Effect: Decoding the Lost Science of Prayer and Prophecy*, (New York, Harmony Books/Random House, Inc., 2000)
159. Ibid

160. Sounds True, Inc., *The Mystery of 2012: Predictions, Prophecies & Possibilities*, edited by Tami Simon, (Boulder, CO, Sounds True, Inc., 2007) citing chapter "A Vision for Humanity" by Barbara Marx Hubbard.
161. Patricia Mercier, *The Maya End Times: A Spiritual Adventure to the Heart of the Maya Prophecies for 2012*, (London, Watkins Publishing, 2008)

CONCLUSIONS

1. Sandra Anne Taylor, *Truth, Triumph, and Transformation: Sorting out the Fact from Fiction in Universal Law*, (Carlsbad, CA, Hay House, Inc., 2010)
2. Gregg Braden, *Awakening to Zero Point: The Collective Initiation*, (Bellevue, WA, Radio Bookstore Press, 1993, 1994, 1997)
3. Mike Dooley, *Infinite Possibilities: The Art of Living Your Dreams*, (New York, Atria Books, A Division of Simon & Schuster, Inc., 2009)
4. Barbara Marciniak with Karen Marciniak and Tera Thomas, *Earth: Pleiadian Keys to the Living Library*, (Sante Fe, NM, Bear & Company, Publishing, 1995)
5. Rev. Deborah L. Johnson, *Letters from the Infinite volume 1: The Sacred Yes*, (Boulder, CO, Sounds True, Inc., 2002, 2006)
6. Patricia Mercier, *The Maya End Times: A Spiritual Adventure to the Heart of the Maya Prophecies for 2012*, (London, Watkins Publishing, 2008)
7. Ibid
8. Gregg Braden, *The Divine Matrix: Bridging Time, Space, Miracles and Belief*, (Carlsbad, CA, Hay House, Inc., 2007)
9. Ibid
10. Ibid
11. Sandra Anne Taylor, *Truth, Triumph, and Transformation: Sorting out the Fact from Fiction in Universal Law*, (Carlsbad, CA, Hay House, Inc., 2010)
12. Gregg Braden, *Awakening to Zero Point: The Collective Initiation*, (Bellevue, WA, Radio Bookstore Press, 1993, 1994, 1997)
13. Whitley Strieber, *The Key: A True Encounter*, (Copyright 2001, 2011 by Walker & Collier, Inc., Published by Jeremy P. Tarcher/Penguin, New York, 2011)
14. Penney Peirce, *Frequency: the Power of Personal Vibration*, (New York, Atria Books, 2009)
15. Gregg Braden, *The Divine Matrix: Bridging Time, Space, Miracles and Belief*, (Carlsbad, CA, Hay House, Inc., 2007)

16. Ibid
17. Ibid
18. Alberto Villoldo, *The Four Insights: Wisdom, Power, and Grace of the Earthkeepers*, (Carlsbad, CA, Hay House, Inc., 2006)
19. Jason Shulman, *The Instruction Manual for Receiving God*, (Boulder, CO, Sounds True, Inc., 2006)
20. Jonathan Goldman, *The Divine Name: The Sound That Can Change the World*, (Carlsbad, CA, Hay House, Inc., 2010)
21. Ibid
22. Alberto Villoldo and Erik Jendresen, *Island of the Sun: Mastering the Inca Medicine Wheel*, (Rochester, VT, Destiny Books, HarperSanFrancisco, a division of HarperCollins Publishers, Inc., 1992, 1995)
23. Mary Ford Grabowsky, *Woman Prayers: Prayers by Women Throughout History and Around the World*, (New York, HarperCollins Publishers, Inc., 2003)
24. As revealed to Rev. Deborah L. Johnson, *Letters from the Infinite volume 1: The Sacred Yes*, (Boulder, CO, Sounds True, Inc., 2002, 2006)
25. Patricia Mercier, *The Maya End Times: A Spiritual Adventure to the Heart of the Maya Prophecies for 2012*, (London, Watkins Publishing, 2008)
26. Ibid
27. Manly P. Hall, *The Secret Teachings of All Ages: An Encyclopedic Outline of Masonic, Hermetic, Qabbalistic and Rosicrucian Symbolical Philosophy*, (New York, Jeremy P. Tarcher/Penguin, 2003)
28. Neale Donald Walsch, *Happier Than God: Turn Ordinary Life into an Extraordinary Experience*, (Ashland, OR, Emnin Books, distributed by Hampton Roads Publishing Company, 2008)
29. Ibid
30. Neale Donald Walsch, *Tomorrow's God: Our Greatest Spiritual Challenge*, (New York, Atria Books, 2004)
31. Neale Donald Walsch, *Mother of Invention: The Legacy of Barbara Marx Hubbard*, (Carlsbad, CA, Hay House, Inc., 2011)
32. Eckhart Tolle, *A New Earth: Awakening to Your Life's Purpose*, (New York, Dutton, Penguin Group (USA) Inc., 2005)
33. Matthew Fox and Rupert Sheldrake, *Natural Grace*, (New York, Doubleday, 1996)
34. Mike Dooley, *Infinite Possibilities: The Art of Living Your Dreams*, (New York, Atria Books, A Division of Simon & Schuster, Inc., 2009)

35. Mary Ford Grabowsky, *Woman Prayers: Prayers by Women Throughout History and Around the World*, (New York, HarperCollins Publishers, Inc., 2003)
36. Caroline Myss, *Defy Gravity: Healing Beyond the Bounds of Reason*, (Carlsbad, CA, Hay House, Inc., 2009)
37. Caroline Myss, *Defy Gravity: Healing Beyond the Bounds of Reason*, (Carlsbad, CA, Hay House, Inc., 2009)
38. Echo L. Bodine, *Echoes of the Soul: The Soul's Journey Beyond the Light Through Life, Death, and Life After Death*, (Novato, CA, New World Library, 1999)
39. Michael Bernard Beckwith, *Spiritual Liberation: Fulfilling Your Soul's Potential*, (New York, Atria Books, A Division of Simon & Schuster, Inc., 2008)
40. Mike Dooley, *Infinite Possibilities: The Art of Living Your Dreams*, (New York, Atria Books, A Division of Simon & Schuster, Inc., 2009)
41. Gregg Braden, *The Isaiah Effect: Decoding the Lost Science of Prayer and Prophecy*, (New York, Harmony Books/Random House, Inc., 2000)
42. Sounds True, Inc., *The Mystery of 2012: Predictions, Prophecies & Possibilities, edited by Tami Simon*, (Boulder, CO, Sounds True, Inc., 2007) citing Gill Edwards, in chapter "Wild Love Sets Us Free"
43. Lynne McTaggart, *The Intention Experiment: Using Your Thoughts to Change Your Life and the World*, (New York, Free Press, A Division of Simon & Schuster, Inc., 2007)
44. Masaru Emoto, translated by Noriko Hosoyamada, *The Shape of Love: Discovering Who We Are, Where We Came From, and Where We are Going*, (New York, Doubleday, 2007)
45. Marc Ian Barasch, *Field Notes on the Compassionate Life: A Search for the Soul of Kindness*, (Rodale, distributed by Holtzbrinck Publishers, 2005)
46. Michael Bernard Beckwith, *Spiritual Liberation: Fulfilling Your Soul's Potential*, (New York, Atria Books, A Division of Simon & Schuster, Inc., 2008)
47. James Redfield, *The Celestine Vision*, (New York, Warner Books, Inc., 1997)
48. James Redfield, Michael Murphy, and Sylvia Timbers, *God and the Evolving Universe*: The Next Step in Personal Evolution, (New York, Jeremy P. Tarcher/Putnam, 2002)
49. Louise L. Hay, *You Can Heal Your Life*, (Carlsbad, CA, (Santa Monica, CA – 1984) Hay House, Inc., 1984, 1987)
50. The New English Translation of the Bible, Oxford University Press, Cambridge University Press, 1961, 1970)

51. Gregg Braden, *Awakening to Zero Point: The Collective Initiation*, (Bellevue, WA, Radio Bookstore Press, 1993, 1994, 1997)
52. Sounds True, Inc., *The Mystery of 2012: Predictions, Prophecies & Possibilities*, edited by Tami Simon, (Boulder, CO, Sounds True, Inc., 2007) citing James O'Dea, in chapter "You Were Born for Such a Time as This"
53. Bill Plotkin, *Soulcraft: Crossing into the Mysteries of Nature and the Psyche*, (Novato, CA, New World Library, 2003) here citing Mary Oliver from "Wild Geese" in *Dream Work* (New York, Atlantic Monthly Press, 1986)
54. Bill Plotkin, *Soulcraft: Crossing into the Mysteries of Nature and the Psyche*, (Novato, CA, New World Library, 2003)
55. Ibid
56. Alberto Villoldo, Ph.D., *Courageous Dreaming: How Shamans Dream the World into Being*, (Carlsbad, CA, Hay House, Inc., 2008)
57. Louise L. Hay, *Heart Thoughts: A Treasury of Inner Wisdom*, Compiled and edited by Linda Carwin Tomchin, (Santa Monica, CA, Hay House, Inc., 1990)
58. Roger Housden, *ten poems to set you free*, (New York, Harmony Books, 2003)
59. Roger Housden, *ten poems to open your heart*, (New York, Harmony Books, 2002) citing Translation by Coleman Barks, *Feeling the Shoulder of the Lion*, (Boston, Threshold/Shambala Publications, 1991)
60. Kim Rosen, *Saved by a Poem*, (Carlsbad, CA, Hay House, Inc., 2009)
61. Daniel Ladinsky, poems translated by Daniel Ladinsky, *Love Poems from God: Twelve Sacred Voices from the East and West*, (New York, Penguin Putnam, Inc., 2002)
62. Norman Fischer, *Opening to You: Zen-Inspired Translations of the Psalms*, (New York, Penguin Group/Penguin Putnam Inc., 2002)
63. Ibid
64. Mary Ford-Grabowsky, *Woman Prayers: Prayers by Women from Throughout History and Around the World*, (New York, Harper-Collins Publishers, Inc., 2003)
65. Edited and with introductions by Monica Furlong, *Women Pray: Voices through the Ages, from Many Faiths, Cultures, and Traditions* (Woodstock, VT, LongHill Partners, Inc., 2001)
66. Esther and Jerry Hicks, The Teachings of Abraham® *Getting Into the Vortex: Guided Meditations CD and User Guide*, (Carlsbad, CA, Hay House, Inc., 2010)

67. Dharma Singh Khalsa, M.D. and Cameron Stauth, *Meditation as Medicine: Activate the Power of Your Natural Healing Force*, (New York, Pocket Books, a division of Simon & Schuster, Inc., 2001)
68. Alberto Villoldo and Erik Jendresen, *Island of the Sun: Mastering the Inca Medicine Wheel*, (Rochester, VT, Destiny Books, HarperSanFrancisco, a division of HarperCollins Publishers, Inc., 1992, 1995)
69. Poems by Nancy Wood, Paintings by Frank Howell, *Spirit Walker*, (New York, Delacorte Press, Bantam Doubleday Dell Publishing Group, Inc., 1993)
70. Alberto Villoldo, Ph.D., *Mending the Past and Healing the Future with Soul Retrieval*, (Carlsbad, CA, Hay House, Inc., 2005)
71. © Interstate Industries, Monroe Products, 1993, Fabor, VA, Lovingston, VA
72. ©1973, CBS Inc./Manufactured by Columbia Records, New York
73. ©R. Carlos Nakai, Canyon Records, Phoenix, AZ, 1968
74. ©Jonathan Goldman & Laraaji, Spirit Music, Inc., Boulder, CO, 2002
75. ©Deuter, New Earth Records, 2008
76. ©Deuter, New Earth Records, 2005
77. ©Richard Warner, ENSO Records, Glendale, WI, 1996
78. Bruce H. Lipton, PH.D. and Steve Bhaerman, *Spontaneous Evolution: Our Positive Future (and a way to get there from here)*, (Carlsbad, CA, New York, Hay House, Inc., 2009)
79. Patricia Mercier, *The Maya End Times: A Spiritual Adventure to the Heart of the Maya Prophecies for 2012*, (London, Watkins Publishing, 2008)
80. quoted by Gregory Sams, in *Sun of gOd:Consciousness and the Self-Organizing Force that Underlies Everything*, (San Francisco, Red Wheel/Weiser, LLC., 2009)
81. John L. Peterson, *A Vision for 2012: Planning for Extraordinary Change*, (Golden, CO, Fulcrum Publishing, 2008)
82. Ibid
83. Ibid
84. Ibid
85. Sounds True, Inc., *The Mystery of 2012: Predictions, Prophecies & Possibilities*, edited by Tami Simon, (Boulder, CO, Sounds True, Inc., 2007) citing Corrine McLaughlin, in chapter "2012: Socially Responsible Business and Nonadversarial Politics"
86. Ibid

87. John Major Jenkins, *The 2012 Story: The Myths, Fallacies, and Truth Behind the Most Intriguing Date in History*, (New York, Jeremy P. Tarcher/Penguin, 2009)
88. Barack Obama, *The Audacity of Hope: Thoughts on Reclaiming the American Dream*, (New York, Crown Publishing Group, a division of Random House, Inc., 2006)
89. Ibid
90. Ibid
91. TUT Enterprises, Inc., *2012 A Wrinkle in Time*, featuring and produced by Mike Dooley, (TUT Enterprises, Inc., 2010)
92. John L. Peterson, *A Vision for 2012: Planning for Extraordinary Change*, (Golden, CO, Fulcrum Publishing, 2008)
93. Ibid
94. http://littlegrandmother.net/TribeofManyColors.aspx
95. http://noosphere.princeton.edu/poethist.html
96. Wayne W. Dyer, *There's a Spiritual Solution to Every Problem*, (New York, HarperCollins Publishers Inc., 2001)
97. Lawrence E. Joseph, *Apocalypse 2012: A Scientific Investigation into Civilization's End*, (New York, Morgan Road Books, an imprint of The Doubleday Broadway Publishing Group, a division of Random House, Inc., 2007)
98. Neale Donald Walsch, *Home With God: In a Life That Never Ends*, (New York, Atria Books, 2006)
99. Ibid
100. www.abraham-hicks.com
101. Jonathan H. Ellerby, Ph.D., *Return to the Sacred: Ancient Pathways to Spiritual Awakening*, (Carlsbad, CA, Hay House, Inc., 2009)
102. Neale Donald Walsch, *Home With God: In a Life That Never Ends*, (New York, Atria Books, 2006)
103. Ibid
104. Esther and Jerry Hicks, (The Teachings of Abraham®) *Getting Into the Vortex: Guided Meditations CD and User Guide*, (Carlsbad, CA, Hay House, Inc., 2010)
105. www.abraham-hicks.com
106. Klaus Heinemann, Ph.D, and Gundi Heinemann, *Orbs: Their Mission and Messages of Hope*, (Carlsbad, CA/New York, Hay House, Inc., 2010)
107. Chris Morton and Ceri Louise Thomas, *The Mystery of the Crystal Skulls: Unlocking the Secrets of the Past, Present, and Future*, (Rochester, VT, Bear & Company, 1997, 1998, 2002)

108. Eckhart Tolle, *Oneness With Life: Inspirational Selections from A New Earth*, (New York, Dutton, Penguin Group (USA) Inc., 2008)
109. Drunvalo Melchizadek, *Serpent of Light: The Movement of Earth's Kundalini and the Rise of the Female Light, 1949 to 2013*,(San Francisco, Red Wheel/Weiser, LLC, 2007)
110. Rhonda Byrne, *The Power*, (New York, Atria Books, 2010)
111. Marci Shimoff with Carol Kline, *Love For No Reason: 7 Steps to Creating a Life of Unconditional Love*, (New York, Free Press, A Division of Simon & Schuster, Inc., 2010)
112. Stephen C. Paul, Ph.D. with paintings by Gary Max Collins, *Illuminations: Visions for Change, Growth and Self-Acceptance*, (New York, HarperCollins Publishers, Inc., 1991)
113. Esther and Jerry Hicks, (The Teachings of Abraham®) *Getting Into the Vortex: Guided Meditations CD and User Guide*, (Carlsbad, CA, Hay House, Inc., 2010)
114. www.abraham-hicks.com
115. Neale Donald Walsch, *Home With God: In a Life That Never Ends*, (New York, Atria Books, 2006)
116. James F. Twyman, *The Moses Code: The Most Powerful Manifestation Tool in the History of the World*, (Carlsbad, CA, Hay House, Inc., 2008)
117. Whitley Strieber, *The Key: A True Encounter*, (Copyright 2001, 2011 by Walker & Collier, Inc., Published by Jeremy P. Tarcher/Penguin, New York, 2011)

CPSIA information can be obtained at www.ICGtesting.com
Printed in the USA
BVOW011229090112

280043BV00001B/66/P